God and the Astronomers

God and the Astronomers

by Robert Jastrow

Afterword by
Dr. John A. O'Keefe
and Professor Steven T. Katz

WARNER BOOKS

A Warner Communications Company

Warner Books, Inc.,
75 Rockefeller Plaza,
New York, N. Y. 10019

 A Warner Communications Company

Printed in the United States of America

First printing: April 1980

10 9 8 7 6 5 4 3 2 1

Library of Congress Cataloging in Publication Data
Jastrow, Robert
    God and the astronomers.
    Reprint of the 1978 ed. published by Norton,
New York; with new afterword.
    Bibliography: p.
    Includes index.
    1. Cosmology.   2. Astronomy—History.   I. Title.
QB981.J27 1980   520   80-10882
ISBN 0-446-97350-5

Acknowledgments

The material in this book was originally developed for the Phi Beta Kappa lecture of the American Association for the Advancement of Science given in Washington on February 14, 1978. Sections are adapted from an article on cosmology published in the *New York Times Magazine*.

I am greatly indebted to Doris Cook for many stimulating discussions relating to the ideas in the book, and for valuable contributions in every aspect of the preparation of the manuscript. Ramona Carpenter and Margaret Neuendorfer provided essential secretarial and editorial support. I should also like to express my appreciation to Sally Bassett for exceptional dedication and ability in carrying out many tasks of editing, indexing, and proofreading.

—R. J.

Contents

1. In the Beginning 1
   Illustrations 6
2. Slipher, de Sitter, and Einstein 13
   Illustrations 19
3. Hubble and Humason 37
   Illustrations 44
   Two Biographies 59
4. The Law of the Expanding Universe 85
5. The Birth and Death of Stars 95
6. The Religion of Science 101
   Epilogue 107
   Supplement: The First Billion Years 115
   Galaxies and Stars in Color 121
   Afterword by Dr. John A. O'Keefe
   *The Theological Impact of the
   New Cosmology* 131
   Afterword by Professor Steven T. Katz
   Judaism, God, and the Astronomers 147
   Sources 165
   Picture Credits 169
   Index 171

In the Beginning

WHEN AN ASTRONOMER WRITES ABOUT GOD, HIS COLLEAGUES ASSUME HE IS either over the hill or going bonkers. In my case it should be understood from the start that I am an agnostic in religious matters. However, I am fascinated by some strange developments going on in astronomy—partly because of their religious implications and partly because of the peculiar reactions of my colleagues.

The essence of the strange developments is that the Universe had, in some sense, a beginning—that it began at a certain moment in time, and under circumstances that seem to make it impossible—not just now, but *ever*—to find out what force or forces brought the world into being at that moment. Was it, as the Bible says, that

"Thou, Lord, in the beginning hast laid the foundations of the earth, and the heavens are the work of thine hands?"

No scientist can answer that question; we can never tell whether the Prime Mover willed the world into being, or the creative agent was one of the familiar forces of physics; for the astronomical evidence proves that the Universe was created twenty billion years ago in a fiery explosion, and in the searing heat of that first moment, all the evidence needed for a scientific study of the cause of the great explosion was melted down and destroyed.

This is the crux of the new story of Genesis. It has been familiar for years as the "Big Bang" theory, and has shared the limelight with other theories, especially the Steady State cosmology; but adverse evidence has led to the abandonment of the Steady State theory by nearly everyone, leaving the Big Bang theory exposed as the only adequate explanation of the facts.

The general scientific picture that leads to

the Big Bang theory is well known. We have been aware for fifty years that we live in an expanding Universe, in which all the galaxies around us are moving apart from us and one another at enormous speeds. The Universe is blowing up before our eyes, as if we are witnessing the aftermath of a gigantic explosion. If we retrace the motions of the outward-moving galaxies backward in time, we find that they all come together, so to speak, fifteen or twenty billion years ago.*

At that time all the matter in the Universe was packed into a dense mass, at temperatures of many trillions of degrees. The dazzling brilliance of the radiation in this dense, hot Universe must have been beyond description. The picture suggests the explosion of a cosmic hydrogen bomb. The instant in which the cosmic bomb exploded marked the birth of the Universe.

Now we see how the astronomical evidence leads to a biblical view of the origin of the world. The details differ, but the essential elements in the astronomical and biblical accounts of Genesis are the same: the chain of events leading to man com-

---

* The exact moment in which this happened is uncertain by several billion years. Because of this uncertainty, I have picked twenty billion years, a round number, as *the* age of the Universe. The important point is not precisely when the cosmic explosion occurred, but that it occurred at a sharply defined instant some billions of years ago.

menced suddenly and sharply at a definite moment in time, in a flash of light and energy.

Some scientists are unhappy with the idea that the world began in this way. Until recently many of my colleagues preferred the Steady State theory, which holds that the Universe had no beginning and is eternal. But the latest evidence makes it almost certain that the Big Bang really did occur many millions of years ago. In 1965 Arno Penzias and Robert Wilson of the Bell Laboratories discovered that the earth is bathed in a faint glow of radiation coming from every direction in the heavens. The measurements showed that the earth itself could not be the origin of this radiation, nor could the radiation come from the direction of the moon, the sun, or any other particular object in the sky. The entire Universe seemed to be the source.

The two physicists were puzzled by their discovery. They were not thinking about the origin of the Universe, and they did not realize that they had stumbled upon the answer to one of the cosmic mysteries. Scientists who believed in the theory of the Big Bang had long asserted that the Universe must have resembled a white-hot fireball in the very first moments after the Big Bang occurred. Gradually, as the Universe expanded and cooled, the fireball would have become less brilliant, but its radiation would have never disappeared entirely. It was the diffuse glow of this ancient radiation, dating back to the birth of the Universe, that Penzias and Wilson apparently discovered.*

No explanation other than the Big Bang has been found for the fireball radiation. The clincher, which has convinced almost the last doubting Thomas, is that the radiation discovered by Penzias and Wilson has exactly the pattern of wavelengths expected for the light and heat produced in a great explosion. Supporters of the Steady State theory have tried desperately to find an alternative explanation, but they have failed. At the present time, the Big Bang theory has no competitors.

Theologians generally are delighted with the proof that the Universe had a beginning, but astronomers are curiously upset. Their reactions provide an interesting demonstration of the response of the scientific mind—supposedly a very objective mind—when evidence uncovered by science itself leads to a conflict with the articles of faith in our profession. It turns out that the scientist behaves the way the rest of us do when our beliefs are in conflict with the evidence. We become irritated, we pretend the conflict does not exist, or we paper it over with meaningless phrases.

---

* Ralph Alpher and Robert Herman predicted the fireball radiation in 1948 but no one paid attention to their prediction. They were ahead of their time.

Ralph Alpher *left* and Robert Herman

ALPHER AND HERMAN. Ralph Alpher and Robert Herman
had predicted the existence of the cosmic fireball in 1948,
while they were working with George Gamow on the "Big
Bang" theory of the creation of the elements.

Alpher and Herman calculated that when the Universe
was young it was very hot, and filled with an intense glow of
radiation that should still be visible today in a weakened
form. If this cosmic fireball radiation could be detected, it
would prove that the Universe began in an explosion.

Sensitive instruments, adequate to detect the remnant
of the cosmic fireball, already existed as a result of radar

George Gamow

work in World War II, but scientists familiar with these instruments either did not know about the work of Alpher and Herman, or did not take it seriously. Herman said recently, "There was no doubt in our minds that we had a very interesting result, but the reaction of the astronomical community ranged from skeptical to hostile."

Later, when the discovery of the fireball radiation turned out to be one of the great scientific events of all time, Alpher and Herman received belated recognition and several prizes from learned societies.

DISCOVERY OF THE COSMIC FIREBALL. In 1965 Arno Penzias and Robert Wilson made one of the greatest discoveries in 500 years of modern astronomy. By accident, they detected the cosmic fireball radiation that Alpher and Herman had predicted. The discovery was made with the large horn antenna visible in the background. The horn, built like an oversized ear trumpet, is sensitive to faint radio whispers that travel through the Universe.

Penzias and Wilson were not looking for clues to the beginning of the world when they made their discovery. While testing their equipment, they noticed an unexplained static coming out of their radio receiver. Looking for its

Arno Penzias *right* and Robert Wilson

cause, they crawled inside the horn and discovered pigeons roosting in the rear. After the pigeons and their litter were removed, the static noise persisted. Apparently the static was not due to a defect in the equipment, but was some kind of radiation from space.

A friend told Penzias about a lecture he had heard on the possibility of finding radiation left over from the fireball that filled the Universe at the beginning of its existence. Penzias and Wilson realized they had detected the fireball. The rest is scientific history.

Penzias *right* and Wilson inside the horn

2

Slipher, de Sitter and Einstein

THE SCIENTIFIC STORY OF GENESIS BEGINS IN 1913, WHEN VESTO MELVIN SLIpher—looking for something else, needless to say—discovered that about a dozen galaxies in our vicinity were moving away from the earth at very high speeds, ranging up to two million miles per hour. Slipher's discovery was the first hint that the Universe was expanding.

Slipher reported his extraordinary finding at a meeting of the American Astronomical Society in Evanston, Illinois in 1914. John Miller, who had been Slipher's professor, was present at the meeting. In 1937 he described the scene to John Hall, now Director of Lowell Observatory at Flagstaff where Slipher made his discovery, and recently Dr. Hall passed the account on to me. Slipher presented his results in a cautious manner and with great modesty, but his slides clearly revealed the tell-tale "red shift," a change in the color of the light from these distant galaxies that indicated, to the trained eye, an enormously rapid motion away from the earth. "Then," said Professor Miller, "something happened which I have never seen before or since at a scientific meeting. Everyone stood up and cheered." Although the assembled astronomers did not know exactly what Slipher's discovery meant, they had a gut feeling that this discovery must be of earth-shaking importance.

One of the people in Slipher's audience was Edwin Hubble, who, as we will see, later picked up Slipher's clues and built them into a new picture of the Universe.

Meanwhile, on the other side of the Atlantic—and by now it was wartime—Einstein published his equations of general relativity in 1917. Willem de Sitter, a Dutch astronomer, found a solution to them almost immediately that predicted an exploding Universe, in which the galaxies of the heavens moved rapidly away from one another. This was just what Slipher had observed.

However, because of the interruption of communications by the war, de Sitter probably did not know about Slipher's observations at that time.

Einstein had failed to notice that his theory predicted an expanding Universe. Later, it turned out that Einstein had missed still another expanding-Universe solution to his own equations. This time the discovery was made by a Russian mathematician, Alexander Friedmann. He found that Einstein had made a schoolboy error in algebra which caused him to overlook the additional solutions. In effect, Einstein had divided by zero at one point in his calculations. This is a no-no in mathematics. As soon as Friedmann corrected the error, the missing solution popped out.

As an aside, Einstein seems to have been quite put out by Friedmann's discovery of his mistake, because in a rare display of discourtesy he ignored Friedmann's letter describing the new solution; and then, when Friedmann published his results in the Zeitschrift für Physik in 1922, Einstein wrote a short note to the Zeitschrift calling Friedmann's result "suspicious," and proving that Friedmann was wrong. In fact, Einstein's proof was wrong.

Friedmann wrote Einstein shortly after Einstein's note appeared in the Zeitschrift, timidly pointing out that the master must have made another mistake. Friedmann was very respectful in his letter to the world-famous scientist, and clearly reluctant to challenge him. Every young person who has quarreled with his senior professor, at great peril to his job, knows the terror that must have been in Friedmann's heart when he wrote, after correcting Einstein's algebra, "Most honored professor, do not hesitate to let me know whether the calculations presented in this letter are correct."

But Friedmann clearly felt that he had discovered something of great importance, and this must have given him courage, for then he went on, mindful of Einstein's initial silence, "I particularly ask you not to delay your answer to this letter," and finally, showing his teeth, "In the case that you find my calculations to be correct . . . you will perhaps submit a correction."

Finally, Einstein acknowledged his double error in a letter to the Zeitschrift in 1923, in which he wrote, "My objection [to the Friedmann letter] rested on an error in calculation. I consider Mr. Friedmann's results to be correct and illuminating." Einstein had accepted the legitimacy of his own brainchild.

Getting back to de Sitter, his theoretical prediction of an expanding Universe made a great impression on astronomers immediately after World War I. For the first time, they saw the larger significance in Slipher's discovery of the outward-moving galaxies. Arthur Eddington, the English astronomer, picked up de Sitter's work and made a big to-do over it. Hubble said later that it was mainly de Sitter's result that had influenced him to take up the study of the moving galaxies where Slipher had left off.

Around this time, signs of irritation began to appear among the scientists. Einstein was the first to complain. He was disturbed by the idea of a Universe that blows up, because it implied that the world had a beginning. In a letter to de Sitter—discovered in a box of old records in Leiden a few years ago—Einstein wrote, "This circumstance [of an expanding Universe] irritates me," and in another letter about the expanding Universe, "To admit such possibilities seems senseless."

This is curiously emotional language for a discussion of some mathematical formulas. I suppose that the idea of a beginning in time annoyed Einstein because of its theological implications. We know he had well-defined feelings about God, but not as the Creator or the Prime Mover. For Einstein, the existence of God was proven by the laws of nature; that is, the fact that there was order in the Universe and man could discover it. When Einstein came to New York in 1921 a rabbi sent him a telegram asking, "Do you believe in God?" and Einstein replied, "I believe in Spinoza's God,

17

who reveals himself in the orderly harmony of what exists."

Returning to the story of the expanding Universe, Slipher continued his labors, collecting the light from ever more distant galaxies and measuring their speeds. A few astronomers made similar measurements and confirmed the accuracy of his results, but for the most part, he worked alone. By 1925 he had clocked the velocities of 42 galaxies. Nearly all were retreating from the earth at high speeds. These accomplishments placed Slipher in the ranks of the small group of men who have, by accident or design, uncovered some element of the Great Plan.

Vesto Melvin Slipher

RETREATING GALAXIES. Vesto Melvin Slipher discovered the retreating motion of the galaxies around 1913. His discovery, like many great findings in science, was accidental. Slipher had been studying the Andromeda nebula at the request of Percival Lowell, director of the Lowell Observatory in Flagstaff. At that time the Andromeda nebula was not known to be a spiral galaxy, and was a mysterious object. Lowell thought this nebula might be a solar system like ours, but still in the process of birth. Since a newly forming solar system is likely to be rotating, Silpher's first task was to look for a rotation, or swirling motion, in the Andromeda galaxy, or nebula as it was known then.

Slipher failed to detect any rotation, but he discovered instead that the entire galaxy was moving relative to the earth at a speed of 700,000 miles per hour. In the following year, Slipher found that about a dozen galaxies within the range of his telescope were moving rapidly away from the earth. He did not realize that he had stumbled upon the first evidence for the expanding Universe.

Percival Lowell

AN HISTORIC MEETING. Slipher reported on the rapid motion of the galaxies in 1914, at a meeting of the American Astronomical Society in Evanston, Illinois. When he was finished, his colleagues rose and applauded him in a spontaneous tribute to the extraordinary nature of his findings.

The photograph *below* records the historic occasion of this meeting. The participants are gathered in front of the Engineering Building on the campus of Northwestern University. The seventy-six Society members who attended

Astronomers at the Evanston meeting

the meeting represented a large fraction of the astronomers living in America at that time. This fact indicates the modest scale of early astronomical research in America.

Edwin Hubble stands in the front row *at far right*. Hubble, then an astronomy student, was elected to membership in the Society at the Evanston meeting. Slipher is at left *in the rear (circle)*. Hubble understood the full significance of Slipher's results; later, he used them, together with his own measurements and those of Humason, to create a new picture of the Universe.

THE KAPTEYN UNIVERSE. This photograph of a group of astronomers at Yerkes Observatory shows Slipher, *far right*, and Jacobus Kapteyn, *facing left* and identifiable as the only bareheaded person in the group. Kapteyn, one of the pioneer explorers of the Galaxy, counted the number of stars visible in various directions in the sky, and concluded that our solar system must be in the center of the Galaxy and may be in the center of the Universe. This idea—known as the Kapteyn Universe—made astronomers, including Kapteyn himself, uneasy.

Today we know the sun is far from the center of our Galaxy. Slipher found the reason for Kapteyn's error: thick clouds of dust, drifting in interstellar space, block the light from the myriad stars near the actual center of the Galaxy, and make it appear that we are at the center. Unfortunately, Slipher's results passed almost unnoticed, and the debate over the sun-centered Universe continued for another 15 years.

Willem de Sitter around 1898

DE SITTER. Willem de Sitter, a Dutch astronomer, played a key role in the sequence of events that established the expanding Universe as an accepted fact. De Sitter was born in the Netherlands in 1872, and studied astronomy in the University at Groningen under Kapteyn. He received his Ph.D. degree in 1897, and spent the next two years in Cape Town, South Africa, observing southern-hemisphere stars. The photograph shows him as a young man in the Cape Town Observatory.

In 1898 de Sitter met Eleanora Suermondt in Cape Town and married her there. They had two boys and two girls. His daughter, Theodora Smit, remembers him as a very good father. "No matter how busy he was, he spent Sunday with the family," Mrs. Smit recalls. He was very friendly and always helped the children with their mathematics lessons.

De Sitter returned to Groningen in 1899, moved to Leiden in 1908, and became director of the Observatory at Leiden in 1919. The photograph *above* shows the de Sitters in the Observatory garden around 1932. De Sitter died in Leiden in 1935.

Willem and Eleanora de Sitter

THEORY OF AN EXPANDING UNIVERSE. Although Slipher's measurements on the moving galaxies implied the Universe was expanding, no one realized this immediately. De Sitter made a theoretical discovery a few years later that brought the concept of the expanding Universe to the forefront of attention.

That story begins in 1916, when Einstein, then in Berlin, sent a copy of his paper on the equations of general relativity to de Sitter in Leiden. De Sitter studied Einstein's equations, and discovered that they had an expanding-Universe solution. Einstein did not like the solution because it implied that the world had an abrupt beginning. However, many astronomers were intrigued by the "de Sitter Universe," in which everything moved away from everything else, and began to think of ways to detect the expansion.

As an aside, de Sitter also set in motion the train of events that catapulted Einstein to world fame. When de Sitter received Einstein's paper on relativity in 1916, he passed it on to Arthur Eddington, a British astronomer. Eddington was a superb mathematician and quickly grasped the essence of the theory. He acclaimed it as "a revolution of thought," and set to work to organize the eclipse expedition that proved the validity of Einstein's ideas in 1919. The expedition measured the bending of light by gravity—an effect predicted by relativity. The dramatic verification of Einstein's theory made him the best-known scientist in the world.

Hendrick Antoon Lorentz was a Dutch physicist whose work provided much of the foundation on which Einstein built the theory of special relativity. Einstein considered Lorentz the greatest scientist he had known. He said about

Lorentz, "He meant more to me personally than anybody else I have met in my lifetime."

The photograph *opposite* was taken in Leiden in 1923, during one of Einstein's frequent visits. Einstein usually stayed with Paul Ehrenfest, one of his closest friends, who shared his relaxed view of academic formalities.

Einstein   de Sitter

                   Ehrenfest

Eddington   Lorentz

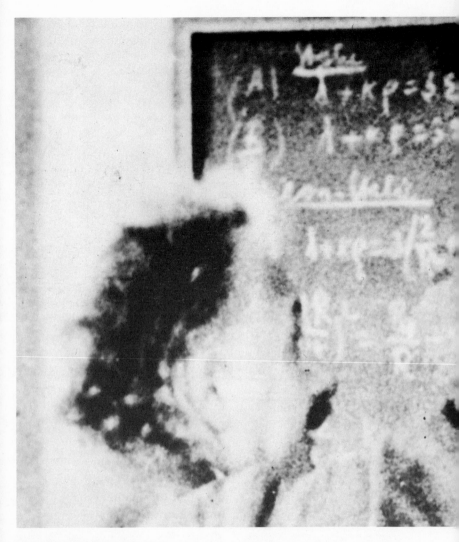

EINSTEIN AND DE SITTER. Einstein met de Sitter during a visit to California in 1932, and discussed de Sitter's theory of the expanding Universe at the blackboard *above*.

Einstein resisted de Sitter's theory for many years after they first corresponded in 1917, but Hubble's observations

on the speeds and distances of the galaxies finally convinced him that the theory was correct. Shortly before his death, he told a visitor that he fully accepted the idea of "a beginning."

EINSTEIN AND LEMAÎTRE. During Einstein's stay in the United States, he met Georges Lemaître *above right* a Belgian priest who studied astronomy under Eddington at Cambridge. In 1927, Lemaître had discovered an expanding-universe solution to Einstein's equations of general relativity.

Lemaître's solution was similar to the one Friedmann had found several years earlier. Friedmann's work had gone unnoticed by astronomers, but Lemaître's theory was publicized by Eddington and became widely known.

3

Hubble and Humason

AFTER 1925, SLIPHER DROPPED THE STUDY OF THE GALAXIES AND TURNED TO other problems. As he left the field, Hubble and Humason entered it, and began to follow up on his work with the large telescopes on Mount Wilson. Slipher himself had never realized the connection between his measurements and the expanding Universe; he had a completely different explanation for the moving galaxies.* Yet he had played a

---

* Slipher believed that the galaxy to which the sun belonged was drifting through space, carrying the sun and earth with it. According to this interpretation, the apparent motions of the other galaxies were only a reflection of our own movement.

crucial role. Many years afterward, Hubble wrote about Slipher's measurements, "The first step in a new field is the great step. Once it is taken, the way is clear and all may follow."

But Hubble saw the connection clearly. He had the golden touch, the knack of working on the important problems. He seems to have been the first American astronomer to understand the relation between Slipher's results and the bold theoretical concepts fashioned by de Sitter out of Einstein's equations. Hubble persuaded Milton Humason to join him in a great undertaking. Working together, they would turn the power of the 100-inch telescope, then the world's largest, on the problem of the moving galaxies.

The techniques for measuring the speeds of the galaxies required infinite patience and care. Humason, a self-taught astronomer who had started out as a mule-train driver and janitor at the Mount Wilson Observatory, was known to his colleagues as a man of "exquisite skill." It was natural for Hubble to turn to Humason for help in the great enterprise. Humason set to work. Many years later, Hubble wrote with affection about their partnership, "Humason assembled spectra of nebulae and I attempted to estimate their distances . . . Humason's adventures were spectacular. He first observed some of Slipher's nebulae, and then, when he was sure of his techniques, and confident of his results, he set forth. From cluster to cluster he marched with giant strides right out to the limit of the 100-inch."

Humason clocked the speeds of many galaxies too distant and faint to be seen by Slipher with his 24-inch instrument. He probed the depths of space out to a distance of more than 100 million light years, and throughout this enormous region all the galaxies he measured confirmed Slipher's discovery; every one was moving away from the earth at a high speed. Some were retreating at the extraordinary speed of 100 million miles an hour.

While Humason measured the speeds of the galaxies, Hubble measured their distances. The distances were the missing pieces in the puzzle. A picture of a galaxy taken through a telescope does not tell how far away it is, because an object that is enormous in size and extremely bright may look small and faint if it is very distant. Were the spiral galaxies large, majestic objects, sailing through the reaches of space? Or were they relatively small and nearby bits of luminous matter? Until astronomers decided between these possibilities for the luminous spirals, they had no hope of deciphering the meaning in their rapid motions.

A few astronomers held the first view; they argued that the spirals* were island universes or

---

* At the time these objects were called "spiral nebulas" because no one knew whether or not they were true galaxies. The term "spiral galaxy" came into use later, largely as a result of Hubble's work.

true galaxies, enormously large and enormously distant, each containing billions of stars. In their opinion, the sun belonged to one island universe of stars among many that dotted the vastness of space. But other astronomers felt uncomfortable with the idea of island universes, which relegated our entire galaxy to an insignificant place in the larger scheme of things. They preferred the second theory, which held that the luminous spirals were small, nearby objects—little pinwheels of gas, swirling in the space between the stars.

Some proponents of this view even argued that each spiral was a newborn solar system, with a star forming in the center of the spiral and a family of planets condensing out of the streamers of gas around it. And James Jeans, a British physicist, thought the spirals were still more mysterious; he suggested that they could be places where matter and energy were pouring into our Universe from some other universe existing in another dimension, like gas escaping from one room into another through a crack in the wall.

Hubble settled the controversy. First, using the 100-inch telescope, he photographed several nearby spiral galaxies with great care, and showed that each one contained enormous numbers of separate stars. His photographs proved that the spirals were indeed island universes, or galaxies, very much like our Galaxy.

Furthermore, since the spiral galaxies contained so many stars they must be very large; yet their apparent size, as seen in the telescope, was quite small. The implication was that they were extremely far away—far outside the boundaries of our Galaxy. This was the first clear indication of the great size of the Universe.

Exactly how far away were the spirals? Hubble thought that if he knew the answer to that question, he could solve the mystery of Slipher's retreating galaxies. He used a simple method for judging distance; in fact, it is the same method used by every person who drives along a narrow road on a dark, moonless night. If a car approaches traveling in the opposite direction, the driver judges how far away it is by the brightness of its headlights. If the lights are bright, the car is close; if they are dim, the car is far away.

Following the same reasoning, Hubble judged the distance to other galaxies by the brightness of the stars they contained. He used the driver's rule of thumb: the fainter the stars in the galaxy, the more distant it was.*

---

* An accurate measurement of galactic distances by this method is complicated by the fact that some stars in a galaxy are much brighter than others. Hubble used a certain kind of star known as a Cepheid variable, whose true brightness was known from the properties of similar stars in our own Galaxy. This method works out to distances of about 10 million light years. Beyond that point, the Cepheid variables in other gal-

In this way, Hubble arrived at values for the distances to about a dozen nearby galaxies. The majority were more than a million light years away, and the distance to the farthest one was seven million light years.*

These distances were staggering; they were far greater than the size of our Galaxy, which is 100,000 light years. A few people had guessed that the Universe is large, but until Hubble made his measurements, no one knew how big a place it really is.

Next, armed with his list of distance measurements, Hubble turned back to Slipher's values for the speeds of these same galaxies, augmented by Humason's more recent observations. He plotted speed against distance on a sheet of graph paper, and arrived at the amazing relationship known as Hubble's law: *the farther away a galaxy is, the faster it moves.* This is the law of the expanding Universe. The same law had been predicted by de Sitter on the basis of Einstein's theory of relativity. The agreement made a tremendous impression on astronomers.

Now both theory and observation pointed to an expanding Universe and a beginning in time.

---

axies are too faint to be seen. For still greater distances Hubble developed other methods, such as using the brightness of the entire galaxy as an indication of its distance.

* A light year is 6 trillion miles.

Still Einstein resisted the new developments and held onto his idea of a static, unchanging Universe until 1930, when he traveled halfway around the world from Berlin to Pasadena to visit Hubble. He studied Hubble's plates, looked through his telescope, and announced himself convinced. He said, "New observations by Hubble and Humason concerning the red shift of light in distant nebulae make it appear likely that the general structure of the Universe is not static."

Around 1930, the model of the expanding Universe derived by Friedmann—and a similar kind of Universe derived by Georges Lemaître—became widely known. Of course, at the same time Hubble published his famous law on the expansion of the Universe. And concurrently there was a great deal of discussion about the fact that the second law of thermodynamics, applied to the Cosmos, indicates the Universe is running down like a clock. If it is running down, there must have been a time when it was fully wound up. Arthur Eddington, the most distinguished British astronomer of his day, wrote, "If our views are right, somewhere between the beginning of time and the present day we must place the winding up of the universe." When that occurred, and Who or what wound up the Universe, were questions that bemused theologians, physicists and astronomers, particularly in the 1920's and 1930's.

Edwin Powell Hubble at the 100-inch telescope

HUBBLE'S MEASUREMENTS OF DISTANCES.

Around 1928, Hubble undertook to determine whether the Universe really was expanding. Before he began his work, the picture of an expanding Universe did not have a firm foundation; astronomers did not believe in it because de Sitter's evidence was entirely theoretical, and Slipher's measurements were too incomplete to be convincing.

Hubble's first step was to find out what the mysterious spiral "nebulas" were. Photographing these objects with the 100-inch telescope, then the world's largest, he found that they contained vast numbers of individual stars. Hubble's photographs convinced astronomers that spiral nebulas were true galaxies, or island universes.

Hubble's next step was to find out how far away the galaxies were. He proceeded to measure their distances, using, as a yardstick, a certain kind of star called a Cepheid, whose true brightness was known from studies of similar stars in our own galaxy. From the degree of faintness of Cepheid stars in other galaxies, he could estimate the distances to these galaxies.

THE ANDROMEDA NEBULA. These photographs of the Andromeda nebula show how Hubble proved that this luminous spiral, and other objects like it in the heavens, were galaxies of stars or island universes. Viewed at low magnification (*opposite*), the nebula appears as a diffuse glow of light. (The spots of light scattered across the photograph are stars in our own Galaxy.) But careful examination of the luminous glow in the photograph reveals a mottled appearance caused by countless separate stars. These individual stars are clearly visible in a detailed photograph of a small portion of the rim of the nebula *(overleaf).*

The bright region in the center of the galaxy is an extremely dense cluster of stars, so close together that they cannot be seen separately even in the largest telescope. These stars are very old, and were the first to form when the galaxy was born.

The two luminous spots above and below the galaxy are small satellite galaxies held captive by the gravitational attraction of Andromeda, as the moon is held captive by the earth. Each satellite galaxy contains several billion stars.

An enlarged view of the rim of the Andromeda Nebula

HUMASON'S MEASURE-MENTS OF SPEED. While Hubble photographed the receding galaxies and measured their distances, Humason measured their speeds. Humason's technique for determining speeds, also used by Slipher, depended on a slight change in color—known as the red shift—in the light emitted from moving galaxies. Humason was exceptionally skilled in handling the giant telescopes during the long time exposures required in the case of very distant and faint galaxies. Some of these galaxies were so faint as to be visible to the eye, even when viewed through a large telescope.

Although the cosmic interpretation of the measurements was due mainly to Hubble, Humason played an essential role in the joint venture that established the law of the expanding Universe. During his career, he measured the speeds of 620 galaxies.

Humason on Mount Wilson

In the photograph, Humason measures the red shift of a galaxy by comparing the colors in its spectrum to those of an ordinary, non-moving source of light. The stage of the microscope holds the half-inch slip of glass on which the spectrum of the galaxy was recorded when the exposure was made at the telescope. A large part of the evidence for the expanding Universe resides in these small pieces of glass.

Humason measuring the red shift of a galaxy

THE MOUNT WILSON TELESCOPE. The 100-inch
telescope on Mount Wilson provided proof that the spiral
nebulas were island universes. This telescope was the

largest in the world in the 1920's, when Hubble did his first
work on the spiral galaxies.

THE 200-INCH TELESCOPE. Hubble and Humason continued their study of the galaxies at the 200-inch telescope on Palomar Mountain. This instrument, completed in 1948, was in its turn the largest instrument in the world until a few years ago, when the USSR completed a telescope with a 236-inch mirror.

HUBBLE IN THE OBSERVER'S CAGE. An astronomer working with the 200-inch telescope often sits in the observer's cage, located inside the long and narrow tube at the top of the structure of girders. The photograph shows Hubble at work in the cage. The view is down the telescope tube. The mirror is visible 55 feet below.

Two Biographies

Edwin Powell Hubble

Edwin Powell Hubble

EDWIN POWELL HUBBLE was an exceptional man among scientists—athlete, scholar, soldier, lawyer and astronomer. Hubble was born in Marshfield, Missouri, November 20, 1889, one of seven children. He won a scholarship to the University of Chicago, studied physics there, was very active in college athletics, and played with the champion basketball team of the West. At one point he boxed the French champion, Carpentier.

In 1910 he graduated from the University of Chicago, and was awarded a Rhodes scholarship. On returning to the United States, he practiced law in Louisville, Kentucky. He "chucked the law for astronomy" and returned to Chicago for graduate work in 1914.

Hubble enlisted in the Army at the start of World War I, was commissioned a Captain, and later became a Major. He was wounded in November 1918, returned to the United States the following year, and went to Pasadena to begin his study of the galaxies.

In 1924, Hubble married Grace Burke in Pasadena. One of Hubble's colleagues at Mount Wilson, W. W. Wright, gave Mrs. Hubble a character sketch of her husband before their marriage: "He is a hard worker. He wants to find out about the Universe; that shows how young he is."

Hubble worked without interruption at Mount Wilson and later at Palomar Mountain, with the exception of a tour of duty at Aberdeen Proving Grounds during World War II. His great achievements with the 100-inch telescope, pushing this instrument to the limit of its range, proved the potential value of still larger telescopes. He worked on the design of the 200-inch telescope and used it from its completion in 1948 until his death in 1953.

Hubble was always sensitive to the larger implications in his results, and their relation to the theories of de Sitter and others, but in his system of values what could be seen through a telescope ranked well above a theoretical idea. The concluding sentence of *The Realm of the Nebulae*, Hubble's classic account of the galaxies, expresses his working philosophy: "Not until the empirical resources are exhausted, need we pass on to the dreamy realms of speculation."

As a college athlete in 1909 *left*

With his sister Lucy in 1917

On Mount Wilson in 1923

With James Jeans and Walter Adams at the Mount Wilson Observatory

Inspecting the 100-inch telescope
with Jeans in 1932

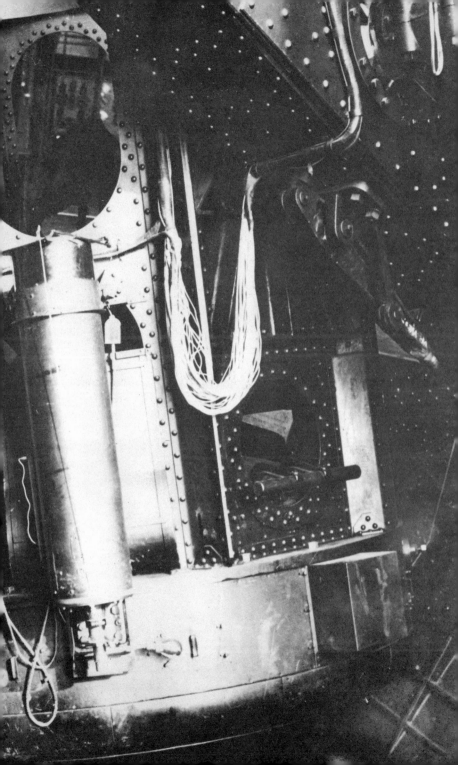

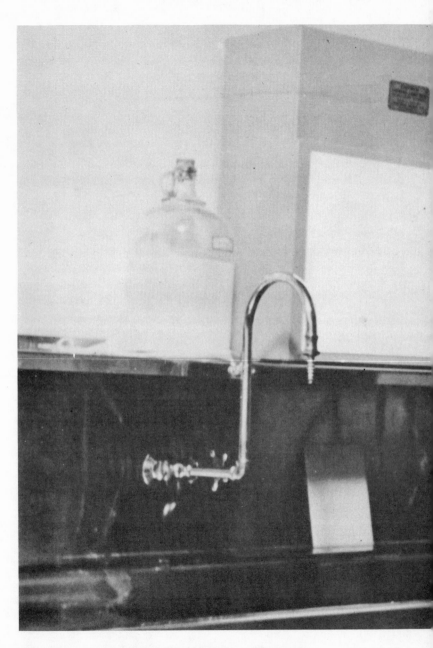

In the darkroom at Palomar Observatory around 1950

Albert Einstein

Albert Einstein

ALBERT EINSTEIN was born in Germany in 1879. He showed no signs of genius as a child, and did rather poorly in school. One of Einstein's teachers told his father, "It doesn't matter what he does; he will never amount to anything."

Einstein attended schools in Germany and Italy, and graduated from engineering school in Zurich in 1901. He had trouble getting a job afterwards. Finally, in 1902, he started to work at the Swiss Patent Office in Bern. The work was to his liking, and he spent his spare time thinking about science. There, in 1905, he wrote the first paper on the theory of relativity. Someone who knew Einstein at the time said later that the years he spent in Bern, virtually unknown, were probably the happiest of his life.

In 1903 Einstein married Mileva Maric, a Serbian physics student. They had two sons. Philipp Frank, who knew the Einsteins a few years later, said that Einstein was very happy with his children, but "life with [Mileva] was not always the source of peace and happiness." The Einsteins were divorced in 1913, and shortly after Einstein married his cousin, Elsa. However, he remained on friendly terms with Mileva and the boys.

By that time, Einstein was established in Berlin as a member of the Prussian Academy of Sciences and a University professor. His colleagues recognized his greatness, and the public lionized him. World fame had come to Einstein overnight in 1919, when a British eclipse expedition reported to the Royal Society that Einstein's new theory of gravity had toppled the 300-year-old theory of Newton. The astronomers had found that Einstein's theory accurately predicted the bending of rays of light by the sun's gravity, which Newton's theory had failed to do. A portrait of Newton looked down on the proceedings, and Alfred Whitehead, the mathematician and philosopher, was present and said later, "The atmosphere of intense interest was exactly that of a Greek drama." From then on, Einstein became the object of a public interest whose intensity was dismaying to him. He wrote to another physicist about the publicity, "[It is] so bad I can hardly breathe."

Einstein's personality contributed to his fame as much as his scientific achievements. His disposition was extraordinarily pleasant and good-humored; Philipp Frank writes of "the laughter that welled up from the very depths of his being." Einstein's manner had an unprofessorial informality; according to Frank, he talked to university officials in the same tone with which he spoke to his grocer, and while most people were pleased by his style, "persons who occupied an important social position" were not so pleased.

During the 1920's, when anti-Semitism was on the rise in Germany, Einstein became the lightning rod for much of the ugliness appearing in German life at that time. He left Berlin for good in 1932, and spent a term in Pasadena as a visiting professor at the California Institute of Technology. In 1933 he came to Princeton as the first member of the staff of the newly established Institute for Advanced Study. Einstein became an American citizen in 1940 by an Act of Congress. He continued to work on problems in theoretical physics in Princeton until his death in 1955.

As a child

As a young man in the Swiss patent office

In 1916, when he published the theory of general relativity

With Elsa, his second wife

In 1932, as visiting professor in California

Taking the oath of citizenship in 1940 with stepdaughter Margot *right* and secretary Helen Dukas

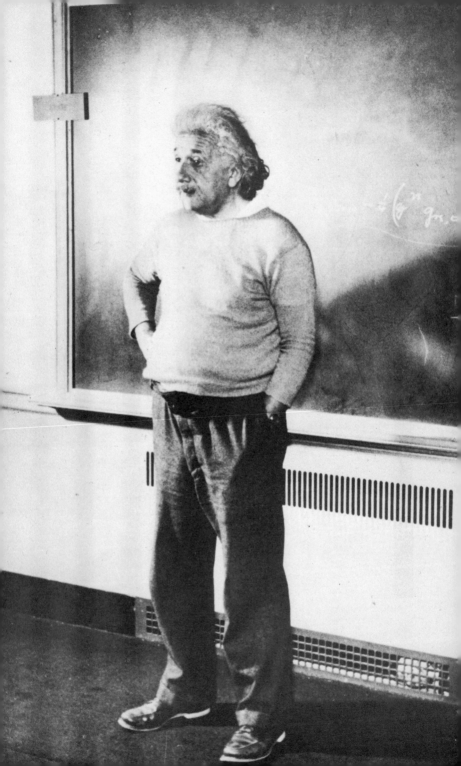

In his later years

At the blackboard in Princeton

The Law of the Expanding

Universe

THE HUBBLE LAW IS ONE OF THE GREAT DISCOVERIES IN SCIENCE; IT IS THE foundation of the scientific story of Genesis. Yet it is a mysterious law. Why should a galaxy recede from us at a higher speed simply because it is farther away?

An analogy will help to make the meaning of the law clear. Consider a lecture hall whose seats are spaced uniformly, so that everyone is separated from his neighbors in front, in back, and to either side by a distance of, say, three feet. Now suppose the hall expands rapidly, doubling its size in a short time. If you are seated in the middle of the hall, you will find that your immediate neighbors have moved away from you and are now at a distance of six feet. However, a person on the other side of the hall, who was originally at a distance from you of, say, 300 feet, is now 600 feet away. In the interval of time in which your close neighbors moved three feet farther away, the person on the other side of the hall increased his distance from you by 300 feet. Clearly, he is receding at a faster speed.

This is the Hubble Law, or the Law of the Expanding Universe. It applies not only to the Cosmos, but also to inflating balloons and loaves of bread rising in the oven. All uniformly expanding objects are governed by this law. If the seats in the lecture hall moved apart in any other way, they would pile up in one part of the hall or another; similarly, if galaxies moved outward in accordance with any law other than Hubble's law, they would pile up in one part of the Universe or another.

One point remains to be explained. How did Slipher and Humason measure the speeds of distant galaxies? It is impossible to make such measurements directly by tracking a galaxy across the sky, because the great distances to these objects render their motions imperceptible when they are observed from night to night, or even from year to year. The closest spiral galaxy to us, Andromeda, would have to be observed for 500 years before it moved a measurable distance across the sky.

The method used by astronomers is indirect, and depends on the fact that when a galaxy moves away from the earth, its color becomes redder than normal.* The degree of the color change is proportional to the speed of the galaxy. This effect is called the red shift. All distant galaxies show a distinct red shift in their color. This fact was first discovered by Slipher. The red shift, which betrays the retreating movements of the galaxies, is the basis for the picture of the expanding Universe.

How is the red shift itself measured? First, a prism or similar device is attached to a telescope.

---

* The effect occurs because light is a train of waves in space. When the source of the light moves away from the observer, the waves are stretched or lengthened by the receding motion. The length of a light wave is perceived by the eye as its color; short waves create the sensation that we call "blue," while long waves create the sensation of "red." Thus, the increase in the length of the light waves coming from a receding object is perceived as a reddening effect.

The prism spreads out the light from the moving galaxy into a band of colors like a rainbow. This band of colors is called a spectrum. In the next step, the spectrum is recorded on a photographic plate. Finally, the spectrum of the galaxy is lined up alongside the spectrum of a nonmoving source of light. The comparison of the two spectra determines the red shift.

The illustration on the facing page shows how the method works. The photographic images of the several galaxies are shown at left, while the spectra of the same galaxies, recorded photographically, appear at right as tapering bands of light. The short, vertical lines above and below each tapering band are the spectrum of a nonmoving source of light, which is placed directly on the photograph for comparison.

The spectra of the galaxies are rather indistinct because the galaxies are faint and far away. However, each spectrum contains one important feature. This is the pair of dark lines circled in white. The lines are colors created by atoms of calcium in the galaxy, which make useful markers for determining the amount of the red shift in a galaxy's spectrum.

The triangle points to the position the calcium colors normally would have in the galaxy's spectrum, if this galaxy were not moving away

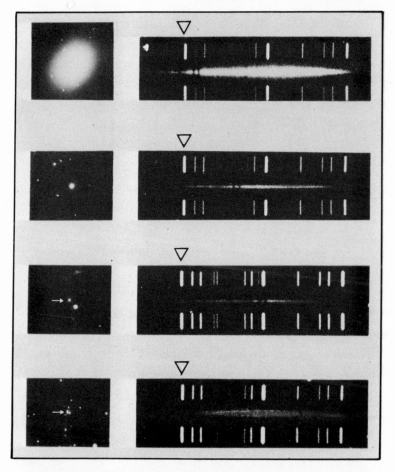

The diffuse spots of light in the photographs *above* are galaxies. The galaxies in the lower photographs, indicated by arrows, are barely visible because they are several billion light years away. The spectrum for each galaxy is the tapering band of light on the right. For each spectrum, the position of the encircled pair of dark lines indicates the amount of the red shift.

from us. The distance between this pointer and the white circle is the amount of the red shift.

The topmost photograph shows a galaxy that is about 70 million light years from us. It is close enough to appear as a large, luminous shape, but too distant for us to see its individual stars. The calcium colors in its spectrum are shifted toward the red by a small but significant amount. The speed of this retreating galaxy, calculated from its red shift, turns out to be three million miles an hour.

The next galaxy is over one billion light years away, and correspondingly smaller and fainter. The position of the calcium colors in its spectrum reveals a much greater shift toward the red, indicating a greater velocity of recession. The red shift in the spectrum of this galaxy corresponds to a speed of 126 million miles an hour.

The third and fourth galaxies are more than two billion light years away. Because of their great distances, they appear as exceedingly small and faint objects. The red shifts in their spectra are very great, and correspond to speeds of recession of more than 200 million miles an hour.

If the speeds and distances of the four galaxies are plotted on a graph, as Hubble plotted similar measurements 40 years ago, the points fall on a straight line.

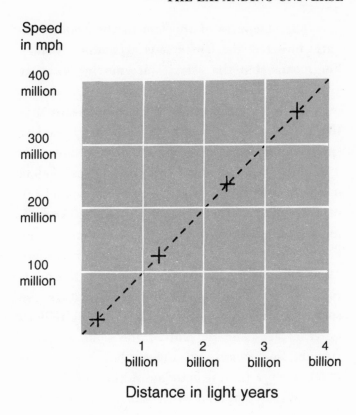

Distance in light years

The line indicates a simple proportion between speed and distance; that is, if one galaxy is twice as far away from us as another, it will be moving away twice as fast; if it is three times as far, it will be moving away three times as fast, and so on. This proportion is the mathematical statement of the Hubble Law.

91

The steepness of the line in the graph indicates how fast the Universe is expanding; a steep line means that the galaxies are moving away at very high speeds; that is, the Universe is expanding rapidly. A line with a gentle slope means that the galaxies are retreating at relatively modest speeds, hence the Universe is expanding slowly.

These remarks about the steepness of the line in the Hubble graph suggest an important check on the theory of the expanding Universe. According to the picture of the explosive birth of the Cosmos, the Universe was expanding much more rapidly immediately after the explosion than it is today. If someone were around to measure the speeds and distances of the galaxies many billions of years ago, and he plotted the same graph, a straight line would still appear, but it would be much steeper than it is today. A copy of that ancient graph, compared with a similar graph today, would test the concept of a Universe that had exploded outward and then slowed down under the pull of gravity.

Can the test be performed? That would seem to be an impossible task, since astronomical records do not go back several billion years. But consider the following facts: the light that reaches the earth from the Andromeda galaxy left that galaxy two million years ago; when an astronomer photographs Andromeda through a telescope, he sees that galaxy as it was two million years earlier, and not as it is today. Similarly, the light that reaches the earth today from the Virgo galaxy left the galaxy 70 million years ago. A photograph of the galaxy shows it as it was 70 million years in the past, and not as it is today.*

Now we see how to obtain a picture of the Universe as it was billions of years ago. First, photograph galaxies that are within a distance of 100 million light years. These galaxies will yield a picture of the expanding Universe as it has been during the last 100 million years. Since 100 million years is a relatively short time on a cosmic time scale, we can consider this picture to represent the Universe as it is today. If the speeds and distances of these relatively nearby galaxies are plotted on a graph, they should form a straight line. The steepness of the line will tell us how fast the Universe is expanding at the present time.

Next, extend the measurements farther out into space, to galaxies whose distances from us are about 500 million light years. The speeds and dis-

---

* This is true in terrestrial affairs also, but the effect is too small to be important. When you see a friend across the room you see him as he was in the past. How far in the past? About one hundred-millionth of a second.

tances of these galaxies will give us another graph, and another line, whose steepness represents the rate of expansion of the Universe approximately 500 million years ago. If the accuracy of our measurements permits us to go still farther out into space, we can measure galaxies at a distance of one billion light years, and then two billion light years, and so on. The farther out we look in space, the farther back we see in time. In this way, using a giant telescope as a time machine, we can discover the conditions in the expanding Universe billions of years ago.

The idea behind the measurement is very simple, but the measurement is hard to carry out in practice because it is difficult to measure the distances to remote galaxies with the necessary accuracy. The most complete study made thus far has been carried out on the 200-inch telescope by Allan Sandage. He compiled information on 42 galaxies, ranging out in space as far as six billion light years from us. His measurements indicate that the Universe was expanding more rapidly in the past than it is today.* This result lends further support to the belief that the Universe exploded into being.

---

* Sandage's work provides additional evidence against the Steady State theory (page 12). If that theory were correct, the rate of expansion would never change.

The Birth and Death of

Stars

ABOUT THIRTY YEARS AGO SCIENCE SOLVED THE MYSTERY OF THE BIRTH AND death of stars, and acquired new evidence that the Universe had a beginning.

According to the story pieced together by astronomers, a star's life begins in swirling mists of hydrogen that surge and eddy through space. The Universe is filled with tenuous clouds of this abundant gas, which makes up 90 per cent of all the matter in the Cosmos. In the random motions of such clouds, atoms sometimes come together by accident to form small, condensed pockets of gas. Stars are born in these accidents (color plates 3 and 4).

Normally the atoms fly apart again in a short time, and the pocket of gas disperses to space. However, each atom exerts a small gravitational attraction on its neighbor, which counters the tendency of the atoms to fly apart. If the number of atoms is sufficiently large, the combined effect of all these separate pulls of gravity will be powerful enough to prevent any of the atoms in the pocket of gas from leaving the pocket and flying out into space again. The pocket becomes a permanent entity, held together by the mutual attraction of all the atoms within it upon one another.

With the passage of time, the continuing attraction of gravity, pulling all the atoms closer together, causes the cloud to contract. The atoms "fall" toward the center of the cloud under the force of gravity; as they fall, they pick up speed and their energy increases. The increase in energy heats the gas and raises its temperature. The shrinking, continuously self-heating ball of gas is an embryonic star.

The ball of gas continues to collapse under the force of its own weight, and the temperature at the center rises further. After 10 million years the temperature has risen to the critical value of 20 million degrees Fahrenheit. At this time, the diameter of the ball has shrunk to one million miles, which is the size of our sun and other typical stars. When the temperature reaches 20 million degrees, a nuclear fire flares up in the center of the star, releasing vast amounts of energy. The release of nuclear energy halts the further collapse of the ball of gas. The energy passes to the surface and is radiated away in the form of heat and light. A new star has been born; another light has appeared in the heavens.

Throughout most of the life of the star, the nuclear fires in its interior burn steadily, consuming hydrogen and leaving behind a residue of heavier elements. These heavier elements are the ashes of the star's fire. Oxygen, iron, copper, and many other elements, ranging up to gold, lead, and uranium, are included among the ashes. According to astronomers, all the elements in the Universe are formed in this way in the interiors of stars, out of the basic building block of hydrogen.

At the end of a star's life, when its reserves of nuclear fuel are exhausted, the star collapses under the force of its own weight. In the case of a small star, the collapse squeezes the entire mass into a volume the size of the earth. Such highly compressed stars, called white dwarfs, have a density of ten tons per cubic inch. Slowly the white dwarf radiates into space the last of its heat and fades into darkness.

A different fate awaits a large star. Its final collapse is a catastrophic event which blows the star apart. The exploding star is called a supernova. Supernovas blaze up with a brilliance many billions of times greater than the brightness of the sun. If the supernova is located nearby in our Galaxy, it appears suddenly as a brilliant new star, visible in the daytime.*

The supernova explosion sprays the material of the star out into space, where it mingles with fresh hydrogen to form a mixture containing all 92 elements. Later in the history of the galaxy, other stars are formed out of clouds of hydrogen which have been enriched by the products of these explosions. The sun is one of these stars; it is a recent arrival in the Cosmos, and contains the debris of countless supernova explosions dating back to the earliest years of our Galaxy. The planets also contain the debris; and the earth, in particular, is composed almost entirely of it. We owe

* Three supernovas visible to the naked eye have been seen in the last 1000 years. The Crab Nebula (color plate 6) is one.

our corporeal existence to events that took place billions of years ago, in stars that lived and died long before the solar system came into being.

This beautiful theory allows the Universe to go on forever in a timeless cycle of death and re-birth, but for one disturbing fact. Fresh hydrogen is the essential ingredient in the plan; it is the main source of the energy by which stars shine, and it is also the source of all the other elements in the Universe. The moment a star is born it be-gins to consume some of the hydrogen in the Uni-verse, and continues to use up hydrogen until it dies. Once hydrogen has been burned within that star and converted to heavier elements, it can never be restored to its original state. Minute by minute and year by year, as hydrogen is used up in stars, the supply of this element in the Universe grows smaller.*

Reflecting on this situation, the astronomer turns the clock back in his imagination and asks

---

* The Steady State theory, which suggests that fresh hydrogen is continually created throughout the Uni-verse out of nothing, avoids this irreversible change, since the freshly created hydrogen can provide the in-gredients for the formation of new stars to replace the old. However, this theory has become untenable be-cause of the discovery of the remnant of the cosmic fireball (pages 12–13).

himself: What was the world like a billion years ago? Clearly there was more hydrogen in the Universe at that time than there is today, and less of the heavier elements. Ten billion years ago, there was still more hydrogen and still less of the heavier elements. Turning the clock back still farther, the astronomer comes to a time when the Universe contained nothing but hydrogen—no carbon, no oxygen, and none of the other elements out of which planets and life are made. This point in time must have marked the beginning of the Universe.

The Religion of Science

NOW THREE LINES OF EVIDENCE—THE MOTIONS OF THE GALAXIES, THE LAWS OF thermodynamics, and the life story of the stars— pointed to one conclusion; all indicated that the Universe had a beginning. A few scientists bit the bullet and dared to ask, "What came before the beginning?" Edmund Whittaker, a British physicist, wrote a book on religion and the new astronomy called *The Beginning and End of the World,* in which he said, "There is no ground for supposing that matter and energy existed before and was suddenly galvanized into action. For what could distinguish that moment from all other moments in eternity?" Whittaker concluded, "It is simpler to postulate creation *ex nihilo*—Divine will constituting Nature from nothingness." Some scientists were even bolder, and asked, "Who was the Prime Mover?" The British theorist, Edward Milne, wrote a mathematical treatise on relativity which concluded by saying, "As to the first cause of the Universe, in the context of expansion, that is left for the reader to insert, but our picture is incomplete without Him."

But the views of most physicists and astronomers were closer to that of Saint Augustine, who, asking himself what God was doing before He made Heaven and Earth, gave the reply, "He was creating Hell for people who asked questions like that." In fact, some prominent scientists began to feel the same irritation over the expanding Universe that Einstein had expressed earlier. Eddington wrote in 1931, "I have no axe to grind in this discussion," but "the notion of a beginning is repugnant to me . . . I simply do not believe that the present order of things started off with a bang . . . the expanding Universe is preposterous . . . incredible . . . *it leaves me cold*." The German chemist,

Walter Nernst, wrote, "To deny the infinite duration of time would be to betray the very foundations of science." More recently, Phillip Morrison of MIT said in a BBC film on cosmology, "I find it hard to accept the Big Bang theory; *I would like to reject it.*" And Allan Sandage of Palomar Observatory, who established the uniformity of the expansion of the Universe out to nearly ten billion light years, said, "It is such a strange conclusion . . . *it cannot really be true.*" (The italics are mine.)

There is a strange ring of feeling and emotion in these reactions. They come from the heart, whereas you would expect the judgments to come from the brain. Why?

I think part of the answer is that scientists cannot bear the thought of a natural phenomenon which cannot be explained, even with unlimited time and money. There is a kind of religion in science; it is the religion of a person who believes there is order and harmony in the Universe. Every event can be explained in a rational way as the product of some previous event; every effect must have its cause; there is no First Cause. Einstein wrote, "The scientist is possessed by the sense of universal causation." This religious faith of the scientist is violated by the discovery that the world had a beginning under conditions in which the known laws of physics are not valid, and as a product of forces or circumstances we cannot discover. When that happens, the scientist has lost control. If he really examined the implications, he would be traumatized. As usual when faced with trauma, the mind reacts by ignoring the implications—in science this is known as "refusing to speculate"—or trivializing the origin of the world by calling it the Big Bang, as if the Universe were a firecracker.

Consider the enormity of the problem. Science has proven that the Universe exploded into being at a certain moment. It asks, What cause produced this effect? Who or what put the matter and energy into the Universe? Was the Universe created out of nothing or was it gathered together out of pre-existing materials? And science cannot answer these questions, because, according to the astronomers, in the first moments of its existence the Universe was compressed to an extraordinary degree, and consumed by the heat of a fire beyond human imagination. The shock of that moment must have destroyed every particle of evidence that could have yielded a clue to the cause of the great explosion. An entire world, rich in structure and history, may have existed before our Universe appeared; but if it did, science cannot tell what kind of world it was. A sound explanation may exist for the explosive birth of our Universe; but if it does, science cannot find out what the explanation is. The scientist's pursuit of the past ends in the moment of creation.

This is an exceedingly strange development, unexpected by all but the theologians. They have always accepted the word of the Bible: In the beginning God created heaven and earth. To which St. Augustine added, "Who can understand this mystery or explain it to others?" The development is unexpected because science has had such extraordinary success in tracing the chain of cause and effect backward in time. We have been able to connect the appearance of man on this planet to the crossing of the threshold of life, the manufacture of the chemical ingredients of life within stars that have long since expired, the formation of those stars out of the primal mists, and the expansion and cooling of the parent cloud of gases out of the cosmic fireball.

Now we would like to pursue that inquiry farther back in time, but the barrier to further progress seems insurmountable. It is not a matter of another year, another decade of work, another measurement, or another theory; at this moment it seems as though science will never be able to raise the curtain on the mystery of creation. For the scientist who has lived by his faith in the power of reason, the story ends like a bad dream. He has scaled the mountains of ignorance; he is about to conquer the highest peak; as he pulls himself over the final rock, he is greeted by a band of theologians who have been sitting there for centuries.

Epilogue

NOW THAT ASTRONOMERS ARE GEN-
ERALLY AGREED ON HOW THE UNIVERSE BE-
gan, what do they have to say about how it will
end? At first thought, it would seem that the Universe must continue to expand forever. As the galaxies fly apart and the distances between them
increase, space grows emptier. Eventually every
galaxy is alone, with no neighbor in view.

Within the isolated galaxies, the old stars
burn out one by one, and fewer and fewer new
stars are formed to replace them. Stars are the

source of the energy by which all beings live. When the light of the last star is extinguished, the Universe fades into darkness, and all life comes to an end.

But many astronomers reject this picture of a dying Universe. They believe that the expansion of the Universe will not continue forever because gravity, pulling back on the outward-moving galaxies, must slow their retreat. If the pull of gravity is sufficiently strong, it may bring the expansion to a halt at some point in the future.

What will happen then? The answer is the crux of this theory. The elements of the Universe, held in a balance between the outward momentum of the primordial explosion and the inward force of gravity, stand momentarily at rest; but after the briefest instant, always drawn together by gravity, they commence to move toward one another. Slowly at first, and then with increasing momentum, the Universe collapses under the relentless pull of gravity. Soon the galaxies of the Cosmos rush toward one another with an inward movement as violent as the outward movement of their expansion when the Universe exploded earlier. After a sufficient time, they come into contact; their gases mix; their atoms are heated by compression; and the Universe returns to the heat and chaos from which it emerged many billions of years ago.

And after that? No one knows. Some astronomers say the Universe will never come out of this collapsed state. Others speculate that the Universe will rebound from the collapse in a new explosion, and experience a new moment of Creation. According to this view, our Universe will be melted down and remade in the caldron of the second Creation. It will become an entirely new world, in which no trace of the existing Universe remains.

In the reborn world, once again the hot, dense materials will expand rapidly outward in a cosmic fireball. Later, gravity will slow down the expansion and turn it into a collapse, followed by still another Creation; and after that, another period of expansion, and another collapse. . . .

This theory envisages a Cosmos that oscillates forever, passing through an infinite number of moments of creation in a never-ending cycle of birth, death and rebirth. It unites the scientific evidence for an explosive moment of creation with the concept of an eternal Universe. It also has the advantage of being able to answer the question: What preceded the explosion?

The answer offered by the oscillating theory is that prior to the explosion the Universe was in a state of increasing density and temperature. As the Universe approached its maximum compression, all the complex elements that had been made within stars during the preceding cycle were melted down, so to speak, into the basic hydrogen out of which they had originally been manufactured. At the moment of maximum compression, another explosion occurred and the Universe was born anew.

How can this theory of an Oscillating Universe be tested? The answer is straightforward. If the density of matter in the Universe is sufficiently great, the gravitational attraction of the different parts of the Universe on one another will be strong enough to bring the expansion to a halt, and reverse it to commence a renewed contraction.* That is, the Universe will be in an oscillating state. On the other hand, if the density of matter in the Universe is not great, the force of gravity will not be sufficient to halt the expansion, and the Universe will continue to expand indefinitely into the future, as predicted by the Big Bang theory.

In other words, the density of matter in the Universe is a critical factor in deciding between the two cosmologies. What is the critical density of matter required to slow down and reverse the

---

* A high density means that on the average, particles in the Universe are relatively close to one another, and therefore, their mutual gravitational attraction is strong.

expansion? A calculation shows that the present expansion of the Universe will be halted if the average density of matter in the Universe corresponds to at least one hydrogen atom in a volume of 10 cubic feet.

How does this threshold value of the density compare with the observed density of matter in the Universe? The matter whose density can be most readily estimated is that which is present in the galaxies in a visible form, as luminous stars and dense concentrations of gas. If we were to smear out the visible matter in the galaxies into a uniform distribution filling the entire Universe, the density of this smeared-out distribution of matter would be too small by a factor of 1000 to halt the expansion.

Since energy is equivalent to matter by Einstein's theory of relativity, we must add to the above figure the contribution from various types of radiant energy in the Universe, such as starlight and the primordial fireball radiation. But these forms of energy turn out to increase the average density of matter by only one or two percent, which is not enough to affect the outcome.

What about matter that is unobservable because it is not luminous? For example, this matter could exist in the galaxies in the form of nonluminous gas in the space between the stars, or as dead stars, or as stars of very low mass and negligible luminosity. It could also be present in the form of gas in the space between the galaxies.

The invisible matter is very difficult to detect, but its amount can be estimated by an indirect method. Galaxies usually are grouped in clusters, the galaxies in a cluster being held together by the force of their mutual gravitational attraction. In such a cluster, the individual galaxies revolve around one another in a swarming motion, like bees in a hive. The more matter a cluster of galaxies contains—in any form, visible or invisible—the stronger the pull of its gravity, and the faster the swarming motions of the galaxies. If the velocities of the galaxies in a cluster can be measured, the total mass of the cluster can be calculated.

This idea has been applied to a large cluster of galaxies called the Coma cluster. The Coma cluster contains 11,000 galaxies—each with billions of stars—packed into a small space with only 300,000 light years separating each galaxy from its neighbors. It is one of the largest organized masses in the Universe.

The results are surprising. On the basis of the motions of the galaxies in the Coma cluster, the amount of matter it contains in an invisible form is roughly thirty times greater than the amount present in the form of luminous stars and other directly observable objects. Yet, although the estimated density of matter in the Universe is greatly increased as a result of this determination, it is still more than ten times too small to bring the expansion of the Universe to a halt.

Thus, the facts indicate that the Universe will expand forever. We still come across pieces of mass here and there in the Universe, and someday we may find the missing matter, but the consensus at the moment is that it will not be found. According to the available evidence, the end will come in darkness.

Supplement: The First Billion Years

WHY DID THE UNIVERSE BEGIN IN AN EXPLOSION? WHAT WERE CONDITIONS LIKE before the explosion? Did the Universe even exist prior to that moment? Most astronomers decline to consider these questions. E. A. Milne wrote, "We can make no propositions about the state of affairs [in the beginning]; in the Divine act of creation God is unobserved and unwitnessed." More recently, James Peebles, of Princeton University, who has made important contributions to the theory of the expanding Universe, said, "What the Universe was like at day minus one, before the big bang, one has no idea. The equations refuse to tell us, I refuse to speculate."

But assuming that some unknown force brought the Universe into being in a hot and highly compressed state, physicists can predict with confidence what happened thereafter. Because of complications introduced by the branch of physics called quantum mechanics, their predictions do not start in the very instant of the explosion—at which time the density was infinite—but only 10^{-43} seconds after that moment, when the density was a finite, but staggering, 10^{90} tons per cubic inch.* At this stage all of the Universe that we can see today was packed into the space of an atomic nucleus. The pressure and temperature were also extremely high, and the Universe was a fiery sea of radiation, from which particles emerged only to fall back, disappearing and reappearing ceaselessly.

The Universe expanded rapidly, and when it was one second old, the density had fallen to the density of water and the temperature had decreased a billion degrees. At this time the fundamental building blocks of matter—electrons, protons, neutrons, and their antimatter counterparts, as well as the ghostlike particles called

---

* There are no names for numbers this large. Written out, it would fill two lines of the page with zeroes.

neutrinos—condensed out of the sea of hot radiation like droplets of molten steel condensing out of the metallic vapor in a furnace.

The Universe continued to expand, and the temperature dropped further. When it fell to around 10 million degrees, protons and neutrons stuck together in groups of four to form helium nuclei. This happened when the Universe was about three minutes old. Calculations indicate that roughly 30 per cent of the hydrogen in the Universe was transformed into helium in this way, in the first three minutes of the Universe's existence.

It might be expected that after helium was formed, other substances would be built up in more complicated nuclear reactions, until all the remaining chemical elements existed. However, the calculations indicate that this does not occur. The reason is that a wide gap exists between helium and the next stable nucleus. By the time helium had been formed, and the next step was about to commence, the temperature and density in the Universe had fallen so low that the gap could not be crossed.

Thus, the theory of the expanding Universe accounts for the presence of an abundance of hydrogen and helium in the Universe, but it fails to explain the existence of carbon, oxygen, iron, gold and all the other chemical elements. Only the story of stellar births and deaths described in Chapter 5 can explain the presence of these substances in the world today.

After the first three minutes, nothing much happened for the next million years. A glow of radiation, left over from the cosmic fireball, pervaded the Universe, obscuring visibility like a thick fog. Particles moved erratically through the fog, colliding with other particles and sometimes with packets of radiant energy.

When the Universe was about one million years old, atoms appeared for the first time. An atom consists of an electron circling in orbit around a nucleus. When the Universe was younger and hotter, any electron captured into an orbit around a nucleus to form an atom was knocked out of its orbit almost immediately, under the smashing impact of the violent collisions that occur at very high temperatures. But by now the temperature had fallen sufficiently so that most electrons could remain in orbit after they were captured. From this moment on, much of the matter in the Universe consisted of atoms.

At the same time the obscuring fog of radiation cleared up, and the Universe suddenly became transparent. The reason for this change was that light, which is a form of radiation, cannot pass through electrically charged particles, such as electrons and protons; however, atoms, which are electrically neutral, do not block radiation appreciably. As soon as the electrons in the Universe had combined with protons or other nuclei to form atoms, rays of light were able to travel great distances unhindered, and it became possible to see from one end of the Universe to the other.*

The void of space has remained transparent down to the present day. This fact has enabled astronomers to see far out into the Universe, and far back into time. In recent years they have photographed quasars—galaxy-like objects of exceptional brilliance—at distances of 15 billion light years. When we see these quasars, we observe the Universe as it was 15 billion years ago, only a few billion years after the beginning cosmic explosion. One might have hoped that with a larger telescope the astronomer could look back to the earliest moments in the life of the Universe. But now we know that because of the obscuring fog of radiation, we will never be able to see anything that happened in the first million years, let alone the first few minutes.

Returning to the main story, with the further passage of time the materials of the expanding Universe cooled and condensed into galaxies, and, within the galaxies, into stars. The galaxies began to form when the Universe was roughly one billion years old. The formation of stars probably began

---

* Of course, no eye was present to perceive the Universe. Neither galaxies, stars, planets nor life existed at that time.

shortly after the formation of the first galaxies. After nearly 20 billion years of continuing expansion, the Universe reached the state in which it exists today.

Galaxies and Stars in Color

PLATE 1. THE WHIRLPOOL GALAXY. The luminous glow in the arms of this spiral is made up of the light from billions of individual stars. Young, hot stars give the arms their blue color. The patch of luminosity below is a small satellite galaxy, held by the gravitational attraction of the multitude of stars in the Whirlpool.

PLATE 2. A PECULIAR GALAXY. This galaxy is one of
the strangest objects in the sky. The spherical glow in the
photograph is the galaxy. The dark lane of matter running
diagonally across its face appears to have been ejected
when an explosion, created by unknown forces, occurred in
the center.

PLATE 3. BIRTH OF STARS: THE TRIFID NEBULA. The glowing masses of the Trifid nebula contain many newborn stars. Hot, young stars, imbedded in the nebula, radiate the energy that creates its beautiful colors. The three dark lines running across the nebula are regions of obscuring dust, which conceal the light of the stars behind them.

124

PLATE 4. BIRTH OF STARS: THE SERPENS NEBULA.
This nebula is also rich in newly forming stars. Each small,
dark region in the photograph is a dense pocket of gas in
the process of becoming a cluster of new stars. The sun
and earth formed out of a dense pocket of gas like these
four and one-half billion years ago.

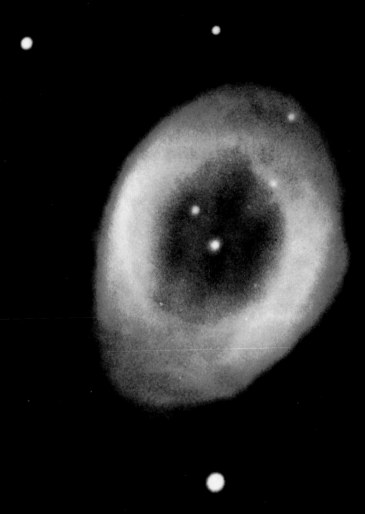

**PLATE 5. DEATH OF STARS: AN AGEING STAR—THE
RING NEBULA.** This expanding shell of luminous gas is a
star close to the end of its life, with its hydrogen fuel nearly
exhausted. In a few billion years, the star will be burned out
and invisible. All small stars suffer this fate. Our sun, a
modest-sized star, will burn out in six billion years.

**PLATE 6. DEATH OF STARS: AN EXPLODED STAR—
THE CRAB NEBULA.** The Crab Nebula is a remnant of a
massive star that came to the end of its life 6000 years ago
and exploded. All massive stars explode when their fuel is
exhausted, dispersing the substance of the star to space.
Later, new stars form out of the remains. The sun, the
earth, and life on the earth are made from the remains of
stars that exploded and died a long time ago.

PLATE 7. THE MILKY WAY: A CROSS-SECTION OF OUR GALAXY. This photograph, showing a minute part of our Galaxy, indicates the multitude of stars in a typical galaxy. Every point of light in the photograph is a sun. Our sun, viewed from a distance, is indistinguishable from one of these points of light. The Universe is made up of billions of galaxies, each containing a comparable multitude of stars.

Afterword

The Theological Impact of the New Cosmology

Dr. John A. O'Keefe, National
Aeronautics and Space Administration

I SHOULD LIKE TO COMMENT AS A PRACTICING CATHOLIC AND AN ASTRONOMER ON the issues raised by the new cosmological discoveries. First let me say, speaking as an astronomer, that I subscribe to Jastrow's view that modern astronomy has found reliable evidence that the Universe was created some fifteen to twenty billion years ago. Jastrow mentions the evidence from the laws of thermodynamics, from the recession of the galaxies, from the calculated length of the lives of the separate stars, and above all, from the cosmic

But even the violent supernova catastrophes do not seem to have obliterated all evidence for a huger and still more ancient event. If we ask how far back in time we have to go in order that the isotopes of uranium and thorium should come into plausible initial relations to one another, there emerges a figure of about nine billion years, as G. Wasserburg has pointed out. We cannot be sure whether this is the date of some great event such as the birth of our Galaxy, or whether it is the average effect of many supernovas throughout the life of the Galaxy; but in any case it points backward to a time much the same as that indicated by the astronomers.

I find it very moving to see how the evidence for the Creation, and even, in some cases, its approximate date, should be so clearly stamped on everything around us: the rocks, the sky, the radio waves, and on the most fundamental laws of physics.

Is the creation which is perceived by contemporary physics the same as the one perceived by ancient and medieval theologians? St. Thomas Aquinas, summing up pre-scientific thought, and deeply influenced by his Jewish, Moslem, and Greek predecessors, had five proofs of the existence of God. The first of these is the argument from motion. Following Aristotle, St. Thomas believed that every moving body must be moved by background radiation, by which we are witnesses of the Big Bang itself.

Another line of evidence is offered by the very matter of the earth itself—the rocks, soil, metal, and wood that we handle daily, and even our own bodies. Among the elements of which matter is made, some, such as uranium, are naturally radioactive. These radioactive substances are made in the bodies of stars during the violent collapse and explosion called a supernova, which terminates the life of nearly every massive star. The explosion spreads the matter of the dying star through space. When the earth first condensed out of interstellar matter, some of its materials had recently passed through a supernova explosion and numerous varieties of radioactive substances were present. Most burned themselves out in a few tens of millions of years. The only radioactive isotopes which survive today in significant quantities are those whose half-lives approach a billion years. Uranium is one such long-lived radioactive element; potassium and thorium are two others. Small amounts of these radioactive substances still exist in the rocks of our planet's crust. Thus the age of the earth is stamped on every pebble; or more exactly, the age of the materials that make up the earth, dated from the moment when these materials were last involved in a supernova catastrophe.

another, hence, by following the chain of motion backward, we should be led to a Prime Mover, moved by no other, "and this, everyone understands to be God."

St. Thomas' physics was incorrect, but it contained the germ of a correct thought. Creating a state of motion demands a supply of energy in an available form.* All the events that occur in the Universe tend to reduce the supply of available energy; this is a loose statement of the second law of thermodynamics. Tracing the chain of events backward in time, we must reach a moment in which all energy is in a state of maximum availability. That moment marks the birth of the universe; it is the starting point for all events that occur thereafter. As Jastrow writes, the Universe was wound up like a clock at this moment, and everything that happened since has been its unwinding. This is the argument for the Creation from the laws of thermodynamics, of which a hint can be seen in St. Thomas' first argument for the existence of God.

His second argument is from the chain of causes. Everything has a cause; if we pursue each effect back to its cause, and that cause to its cause, we will eventually come, says St. Thomas,

---

*"Free energy" to the physicist.

to a First Cause, for which no cause can be given and "to which everyone gives the name of God." This is the program of cosmology; it has been done, and as anticipated, it does lead to the Creation and the Creator.

The third and fourth arguments are philosophical, and not related to science. The fifth argument is the argument from design; i.e., the argument that the whole Universe is directed toward some purpose, and that the design is evidence of a designer. I would like to discuss this point in some detail.

I think there may be theologians of any religion who will ask whether the Creator of this stupendous Universe consisting of hundreds of billions of galaxies, each galaxy containing hundreds of billions of stars, can possibly be interested in mankind. We are such a small phenomenon, on the surface of a planet which is only 1/300,000 of the mass of just one of the stars. As the Hebrew Psalmist said, three thousand years ago:

> *When I consider the heavens, the work of*
> *thy hands,*
> *And the moon and the stars which thou*
> *hast made*
> *What is Man, that thou art mindful of*
> *him*
> *And the son of Man, that thou visitest*
> *him?*

AFTERWORD

Is it credible that God, who made these gigantic and appalling wastes of space, really cares for us?

John A. Wheeler, at Princeton, has an interesting scientific argument which seems to say that there is a connection between the creation of the Universe and the mind of man.* Wheeler draws attention to the fact that the forces of gravity, and the inertial forces responsible for the expansion of the Universe, are closely balanced even now, after billions of years of expansion. Galaxies are places where, locally, the expansion of the Universe has been halted by the mutual gravitation of matter. If the rate of expansion had been a little greater at the beginning, then the Universe would have expanded forever. There would have been no galaxies, and therefore no stars, planets, or life.

On the other hand, if the explosion had been a little less violent and the initial rate of expansion a little less rapid, the Universe would have collapsed in a short time—say a few million years, or even a few minutes. In that brief time, evolution could not have produced intelligent life. For instance, two billion years were taken up on the

---

*Sir Bernard Lovell reaches a similar conclusion in *In the Center of Immensities* (New York: Harper and Row, 1978), pp. 122–125.—R.J.

earth by life rising from the blue-green algae to the amoeba. Neither the rapidly expanding Universe nor the slowly expanding Universe could have yielded intelligent beings. Wheeler found that in its early stages the Universe was balanced on a knife edge between these two destinies.

How did the Universe come to be made in this very precise way? Why does it seem as though the Universe was designed for life and for man?

One thought is that there are billions of other universes, which have expanded either too fast or too slowly. All these are necessarily sterile, and we inhabit the only universe which we could possibly inhabit. Wheeler's colleague, Robert Dicke, has gone on to ask what sense there can be in talking about universes which are completely and forever undiscoverable. Wheeler suggests that they do not exist; that the existence of a universe is somehow tied up with the presence, sooner or later, of intelligent observers in it. He thus sees a clear connection between the creation of our Universe and the mind of the intelligent beings in it.

It is not only the rate of expansion of the Universe that must be delicately adjusted if there is to be intelligent life. For instance, Wheeler remarks that if the fine-structure constant, which is the charge on the electron, e, squared and divided by the modified Planck constant $\hbar$ times the velocity of light, c, i.e.,

$$\frac{e^2}{\hbar c}$$

has a value much different from the actual value (1/137, in any units), then stars will be either too dim to give light of the kind needed for life, or else will rush through their lives so fast that life cannot evolve. In this and other ways, Wheeler believes that the relative values of the great constants of physics are determined. They have the values that they do, because if they did not, we would not be here.

> *Consider the lilies of the field, how they grow; they toil not, neither do they spin.*
> *Yet I say unto you that even Solomon in all his glory was not arrayed like one of these.*
> *Wherefore if God so clothe the grass of the field, which today is, and tomorrow is cast into the oven, shall he not much more clothe you, O ye of little faith?*
> *(Matthew 6:28–30)*

Among biologists, the feeling has been since Darwin that all of the intricate craftsmanship of life is an accident, which arose because of the operation of natural selection on the chemicals of the earth's shell. This is quite true; but to the astronomer, the earth is a very sheltered and protected place. There is a marvelous picture from Apollo 8 of the blue and cloud-wrapped earth, seen just at the horizon of the black, cratered, torn and smashed lunar landscape. The contrast would not be lost on any creature; the thought "God loves those people" cannot be resisted. Yet the moon is a friendly place compared to Venus, where, from skies forty kilometers high a rain of concentrated sulfuric acid falls toward a surface that is as hot as boiling lead. Even this is friendly and homelike by comparison with the trillions of kilometers of hard vacuum which separate the stars, or the million-degree temperature of stellar matter, or the tons per cubic centimeter of white dwarfs, or the unspeakable horror of neutron stars and black holes, where normal matter shrieks in agony as it is drawn into that pit from which nothing comes back.

Even these are normal, honorable, comprehensible things compared to what would happen if the physical constants were just a little different. Then nothing would exist except gases, either compressed so dreadfully that a whole universe would occupy the volume of a pinhead, or spread out in a tenuous cloud, thinner than the best laboratory vacuum, and extending to infinity.

We are, by astronomical standards, a pampered, cossetted, cherished group of creatures; our Darwinian claim to have done it all ourselves is as ridiculous and as charming as a baby's brave efforts to stand on his own feet and refuse his mother's hand. If the Universe had not been made with the most exacting precision we could never have come into existence. It is my view that these circumstances indicate the Universe was created for man to live in.

We see, then, that the resemblance between our cosmology today and that of the theologians of the past is not merely accidental. What they saw dimly, we see more clearly, with the advantage of better physics and astronomy. But we are looking at the same God, the Creator.

What will all this do to our theological ideas in the Catholic Church?

Nothing but good, I believe. As I see it, there is no use telling people that they ought to believe in God. We cannot believe what we do not think is true. We have to have assurance that someone has looked at the evidence and satisfied himself that that is what it means. We do not have to do it ourselves; most of us are not equipped to do it. But even very simple people sometimes make remarkably good choices of the people to trust. I think that the confirmation that the Universe was created at a definite time in the past, and that we see no reasonable prospect of explaining the Creation in natural terms, will be seen by many people as a starting-point for faith.

It may also make Catholics look again at their Scholastic and Jesuit heritages. The effort which these scholars made to relate religion to science was immensely fruitful for both. On the one hand, the historian of science must be well aware that it is often the branches of science which seem to have the greatest theological impact that are most rapidly developed (astronomy at all times, geology in the late nineteenth century, physics in the twentieth century). Pascal, Descartes, Newton, Leibniz, Darwin, Pasteur, Kelvin, Lyell, Einstein, Schrodinger, Heisenberg, Eddington, and Jeans were all involved in theology as well as science. On the other hand, no one today reads the Bible in the literal way that some post-Reformation churchmen did. It was the scientists who first showed us that Genesis could not be read as scientific cosmology.

Pope Pius XII perceived clearly the value of the new cosmology for religion. He spoke about it in his allocution to the Pontifical Academy of Science in 1951:

> *In fact, it would seem that present-day science, with one sweeping step back across millions of centuries, has succeeded in bearing witness to the primordial* Fiat lux *uttered at the moment when, along with matter, there burst forth from nothing a sea of light and radiation, while the*

> *particles of chemical elements split and*
> *formed into millions of galaxies.*

He went on to say that science has

> *followed the course and direction of cos-*
> *mic developments and just as it was able*
> *to get a glimpse of the term towards which*
> *these developments were inexorably lead-*
> *ing, so also has it located too their begin-*
> *ning in time some five milliard years ago.*
> *Thus, with the conclusiveness which is*
> *characteristic of physical proofs, it has*
> *confirmed the contingency of the universe*
> *and also the well-founded deduction as to*
> *the epoch when the cosmos came from the*
> *Hands of the Creator. Hence creation took*
> *place in time. Therefore there is a creator.*
> *Therefore God exists.*

Other conclusions of theological interest follow from the astronomical evidence for the creation of the Universe, when combined with
evidence bearing on the age of the earth. While
the Universe was created fifteen or twenty billion
years ago, the earth is, according to planetary scientists, only 4.6 billion years old. Therefore, man
and his planet are recent arrivals in the cosmos.
Innumerable planets were created before the
earth, and may bear intelligent life of an age and
wisdom matching or surpassing our own. Is there

any room in this vast Universe of intelligent beings for the belief that God has chosen our planet
to be the sole or even the primary object of His
concern?

The validity of the question depends upon
our acceptance of the notion that intelligent life is
common in the Universe. For my part, I am not so
sure that intelligent life exists on other planets.
The basic argument for this view is that each star
offers life an opportunity, and there are 10^{22} (ten
thousand million million million) stars and planets
in the observable universe. Even if the chance of
life evolving is as small as, say, one in a million,
still there must be millions upon millions of inhab-
ited planets in the Universe.

Suppose, however, that twenty-two separate
conditions must be met for intelligent life: the star
must be single, it must produce visible and ultra-
violet light; its planet must have an atmosphere
that transmits light but not X rays or extreme ul-
traviolet; there must be liquid water; there must
be carbon; the star must live a long time; its out-
put of energy must not vary rapidly; the planet
must be in a suitable zone of distances from its
star; it must have land as well as water; it must
not suffer excessive and prolonged bombardment
by meteorites; and so on.

These conditions would not be satisfied on
every planet in the Universe. If each were satisfied

on only 1 planet in 10, which is not an unreason-
able estimate, then if the requirements are really
separate, the chance of finding a planet with all
22 conditions satisfied simultaneously would be
one tenth multiplied by itself twenty-two times, or
$1/10^{22}$. This would mean that only one planet in
the Universe is likely to bear intelligent life. We
know of one—the earth—but it is not certain that
there are many others, and perhaps there are no
others.

If it should turn out that other planets bear
intelligent life, then certain theological questions
would be raised: e.g., did God also send his Son to
them, or is it our job to evangelize them? These issues were raised in the eighth century by the Irish
missionary Vergilius. Vergilius had deduced from
the sphericity of the earth the fact that there
must be a race of men who lived on the other
side—the Antipodes. These, it was clear, had not
heard of Christ; what should we think about their
salvation? This problem was solved by active missionary work in America and Australia, as these
new continents became accessible.

A greater challenge would be presented by
the discovery of a nonhuman race of intelligent
beings, because of the notion of original sin. We
human beings inherited original sin from our ancestors. Clearly, this new race, independently
evolved on another star, could not inherit the sin

of *our* ancestors. Would they have original sin at all? Christ came to us as the Redeemer for our sins; what about theirs? But Catholic theologians see nothing contrary to faith in the view that God has included all races in the Universe in the saving power of Christ.

Thus I think that the theological impact of modern cosmology lies, not in the possibility of life on other worlds but in the evidence for the creation of the Universe, and the evidence of a design acting through it.

It seems to me that efforts such as Jastrow's to compare the results of physical investigations with those of theology are of great value. First of all, for the young person whose faith is facing its baptism of fire, there is a great need for solid facts that the other side will respect. When I was fourteen, I went away to school, and promptly started to argue about God with my roommate; it was those discussions which turned my mind to astronomy. Jeans was then writing his books on cosmogony; and it soon became clear that this line of argument was the credible one to use.

Second, I think that at any age, one needs to be able to imagine the Creation in some way related to the images and ideas of one's own time. The writer of Genesis lived in a place where clay was ubiquitous; he describes God making man out of clay. He uses images of serpents, orchards, and

swords which were familiar to his time. It is difficult for people now to see the essential underlying
truths through these pictures from another period; it is an immense help to hear these truths re-
framed in the ideas of our own times, and with
images related to the galaxies, the photons, and
the electrons which one hears about in the news-
papers or uses in a television set.

At a higher level, I think that these discussions may set a pattern for the interpretation of
the Scripture. They show us that we must neither
ignore the Scriptural message nor accept it as literal science. Jastrow's respectful treatment of the
theologians will, I hope, help us both to listen to
the other side.

Judaism, God, and the Astronomers

Professor Steven T. Katz,
Department of Religion, Dartmouth
College

THE SCIENTIFIC REVOLUTIONS OF THE PAST THREE CENTURIES HAVE CHALlenged the traditional world views of the major religions of mankind. The dialogue is best known through the interaction between Christian thought and science, as in the conflict between Galileo and the Papacy in the seventeenth century, and between Darwinism and nineteenth-century religious beliefs in the biblical account of creation. But modern science poses profound questions for Judaism[1]* as well.

---

*References to superior numbers will be found on pages 162–163.

Basic to the discussion of Judaism and scientific cosmology is the fact that in the Jewish religion "the deed is the essential thing." Hence, the greatest intellectual efforts of the Jewish tradition have been spent on understanding and clarifying the Torah[2] to extract from it rules of behavior, both "duties of the body" and "duties of the heart." As a consequence, Judaism is more an *orthopraxis*, or religion emphasizing correct behavior, both inner and outer, than an *orthodoxy*, or religion emphasizing correct beliefs. Doctrines and beliefs are indeed integral to Judaism, e.g., belief in a strict monotheism, or belief in the divinely revealed origin of the Torah, but their role in Jewish religious thought, although central, is limited. As a consequence, Judaism permits considerable freedom in the realm of ideas. Thus, for example, allowing for what Genesis tells us, Judaism is open to many interpretations and differences of opinion on just what Genesis means. Indeed, it is probably true to say that there is no one correct Jewish answer to such questions as the "how" of Creation. Certainly some opinions are incompatible with Judaism, and majority and minority views exist within the traditional sources, e.g., the Mishna, the Gemara, the medieval and modern commentaries and codes,[3] but no systematic attempt has been made over the centuries to define an orthodox cosmology to which every Jew must subscribe, beyond the affirmation that the world was brought into being "somehow" by God.[4]

Another fact to be borne in mind is that Judaism is not a fundamentalist religion; Jewish religious tradition does not propose to be carrying out the word of God as revealed in the Bible, without human interpretation. The basic assumption of rabbinic Judaism is that while the Torah is the literal revelation of God to Moses at Sinai, and eternally valid for all generations, it requires interpretation (Deut. 17:11). It is made explicit by the Sages of the Talmud that the Torah provides broad and general regulations, while the process of extracting the full significance of these prescriptions, with all relevant details and corollaries, is left to human reason guided by tradition. For example, the Torah speaks of marriage, but does not specify what constitutes marriage; or again, it forbids "work" on the Sabbath, but fails to specify what constitutes work. Is lighting a lamp work? Is cooking work? If Israel were to have legitimate marriages and refrain from desecration of the Sabbath, it had to "interpret" the implications of these Divine Commands. Thus within agreed limits, and using agreed procedures, mankind is free, and even encouraged from necessity, to search out the meaning of Torah.

Through the historic desire of countless generations of the Jewish people to be guided by the Torah, this process of interpretation was constantly called into play to renew continually the significance of God's revelation in the midst of new or changing circumstances. This process of explication and exegesis is known in Judaism as the Oral Torah (*Torah she be-al Pe*) and is the legitimate, as well as necessary, companion of the Written Torah (*Torah she-bichtav*).

The interpretation of the Written Torah is a complex matter of the most fundamental religious significance. Hence, rules of biblical interpretation, as well as more general theological-hermeneutical principles, needed to be agreed upon by the Sages, for without common rules of procedure there could be no agreed interpretations of Scripture, and thus no valid substantive conclusions. As a consequence, in this fundamental sense Judaism is a "method" as well as a set of teachings and laws.

The significance of this theory of the necessity of biblical interpretation for the encounter of scientific claims and Judaism is that it legitimates interpretative moves that might lessen any tension existing between Scripture and science by, for example, reading certain passages of Scripture allegorically or metaphorically. Thus, Maimonides, the greatest of medieval Jewish thinkers, felt free to write regarding the understanding of the secrets of creation (*Maaseh Bereshit*) that they have "been treated (in Genesis) in metaphors, in order that the uneducated may comprehend it according to the measure of their faculties and the feebleness of their comprehension; while the educated take it in a different [i.e., allegorical or nonliteral] sense".[5] Maimonides' remarks provide the appropriate introduction to Judaic discussion of the specifics of the new cosmology in relation to Jewish thought.

Creation

Cosmologists have debated the "Big Bang" theory of cosmic origin versus the steady theory of an eternal Universe for half a century. Now the matter appears to be settled to the satisfaction of the majority of astronomers in favor of the Big Bang, a scientific version of the Creation. But then one asks, did the Creation occur *ex nihilo*, out of nothing, or was our Universe formed out of pre-existing matter? My understanding of the scientific debate is that astronomers have forsworn any opinion on this question. In Jewish thought, as in Christian thought, the majority of traditional authorities reject such arguments as those of Aristotle for the eternal existence of matter in favor of creation *ex nihilo*. The *Midrash on Genesis* (1:9) records the following encounter:

> *A certain philosopher asked R. Ga-*
> *maliel, saying to him: "Your God was in-*
> *deed a great artist, but surely He found*
> *good materials which assisted Him?"*
>
> *"What are they?" said [R. Gamaliel]*
> *to him. "Tohu, bohu (darkness, water),*
> *ruah (wind), and tehom (the deep)," re-*
> *plied the philosopher.*
>
> *"Woe to that man,' [R. Gameliel] ex-*
> *claimed. "The term 'creation' is used by*
> *Scripture in connection with all of them."*
> *[i.e., they are all created by God and are*
> *not co-eternal with Him]:* Tohu *and*
> bohu: "I make peace and create evil"
> *(Isa. 45:7); darkness:* I form the light,
> and create darkness *(ib.); water:* Praise
> Him, ye heavens of heavens, and ye wa-
> ters that are above the heavens *(Ps.*
> *148:4);* For He commanded, and they
> were created *(ib. 5); wind:* For, lo, He
> that formeth the mountains, and crea-
> teth the wind *(Amos 4:13); the depths:*
> When there were no depths, I was
> brought forth *(Prov. 8:24).*[6]

Here Aristotle's view of the eternity of matter is
rejected by the Sages of the Talmudic era in favor
of creation, and specifically creation *ex nihilo.*
Still again in the medieval era, Maimonides, in the
Guide for the Perplexed, gives both a paradigmatic
Jewish statement on the matter, and teaches a

fundamental methodological lesson regarding the encounter of science and Judaism. "In my opinion," he writes,

> *none of what Aristotle and his followers adduce in support of the world's eternity is a conclusive demonstration. Rather there are grave doubts surrounding their proofs, as you shall shortly hear.*
>
> *What I hope to establish is that the world's coming into being, the doctrine of our Law, which I have explained, is not impossible, and that all the Philosophers' arguments to the effect can be refuted . . . If I can accomplish this and show that both creation and eternity are possible, then it will be possible, I believe, with the question reopened, to receive an answer from revelation, which makes clear things which thinking alone has not the power to reach. . . .*
>
> *Once I have made it clear that what we believe is possible, I shall undertake to show that it is the most probable of the contending views as well, using arguments based on reason: i.e., I shall endeavor to show the preferability of the doctrine of the world's creation; I shall show that even more embarrassing consequences follow from the doctrine of the world's eternity.* [7]

In other words, when the answer is not determined by scientific evidence, the Jew chooses the biblical over the non-biblical position because of the further authority of Scripture. Conversely, the implications of Maimonides' argument are clear: had Aristotle been indubitably correct, he would have accepted Aristotle's metaphysics and interpreted Genesis accordingly.

The Jewish response in favor of creation *ex nihilo* is buttressed by a compelling and independent philosophical argument, which is also employed by the advocates of the Big Bang cosmology. The argument is simple to state: whatever conditions govern the processes of the existing world, these conditions need not apply either to a pre-creation reality, i.e., God in His pristine Wisdom and Estate, or to the instant fact of creation itself. For God creates the world, and the laws of nature which govern therein, without being Himself bound by these laws. In this sense Judaism and the astronomers move to somewhat parallel positions, namely, that the circumstances operative in the instant of creation cannot be deduced from later events.

From the scientific point of view one speaks of the meltdown, in the intense heat of the "Big Bang," of all the evidence bearing on the cause of the explosion. Hence, the scientific account precludes all possibility of reconstituting the pre-creation state and limits scientific investigation to the post-creation reality. From the Jewish point of view one speaks of God as Creator of nature, not nature's servant, hence free of its causal and physical determinants. For both, the Creation is a unique event whose character transcends the application of physical theories and laws valid in the created world. From *Bereshit Rabbah* (1:10):

> *Bar Kappara quoted [the biblical verse]: "For ask now of the days past which were before these, since the day that God created man upon the earth. (Deut. 4:32)": [and commented thereon] you may speculate from the day that days were created, but you may not speculate on what was before that.*

So it must be, for the notion of creation does not properly belong to the scientific vocabulary, which deals in causal connections and is premised on the assumption that causality operates everywhere and over everything. Whether or not the Big Bang cosmology complements or parallels the Genesis account, it does reinforce an overriding consideration: to talk of creation is to point to another category of reality, requiring at least an openness to other than narrowly scientific questions, and even more important, an openness to other than narrowly positivistic answers.

Evolution

Professor Jastrow remarks that while science has no explanation for the birth of the Universe, it has a fairly complete account of the events that have taken place since that moment, leading from the Creation to man. This account, which seems to place the origin of man in the animal world and the world of inanimate matter, receives its strongest support from Darwin's theory of evolution.

Much of the religious thought of the latter part of the nineteenth century was taken up with questions and difficulties posed for classical theism by Darwin's theory. Jewish thinkers, too, became engaged in a debate on evolution that still continues. Although specific details of evolutionary theory are still subject to revision, with new evidence regularly calling older assumptions into doubt, the reality of some form of long-continued evolutionary process seems certain. As a consequence, Jewish religious intellectuals must accept the fact of evolution and measure its implications.

Most of the classical Jewish sources relevant to the discussion are *aggadic*[8] in character, i.e., nonbinding opinions of the Sages which reflect their contemporary world-view; they are not *halachic* rulings, i.e., classical religious-legal prescriptions that must be obeyed. Thus, later generations are free to accept or reject them in the light of the ideologies and scientific wisdom of their own age. It is important to recognize the nonlegal status of these opinions; because of it, the classical Jewish texts which treat such subjects as the origin of man are not Jewish equivalents to dogmas of faith as this term is understood in Christian theology. Thus, disagreement, reconsideration of, and acceptance or rejection of their conclusions are legitimate postures that are theologically permissible.

The majority of classical Jewish sources which describe man's creation do so in ways which can be labeled nonevolutionary. Yet, even among the Talmudic Sages one finds opinions that lend themselves to an evolutionary doctrine. For example, in the Avot d'Rabbi Nathan the following account of man's creation is given:

> *How was Adam created? In the first hour the dust of which he was made was collected; in the second the model after which he was formed was created; in the third the soulless lump of him was made; in the fourth his limbs were tied together; in the fifth orifices were opened in him; in the sixth a soul was added to him; in the seventh he stood upright on his feet; in the eighth Eve was joined to him; in the ninth*

*he was brought into the Garden of Eden;
in the tenth he was commanded; in the
eleventh he sinned; in the twelfth he was
banished and made to leave the garden.* [9]

More recently, Rabbi Norman Lamm, now
president of Yeshiva University in New York
City, made the following exegetical suggestion intended to reconcile evolution and Genesis:

> ... *after the first moment of creation* ex
> nihilo, *when the formless primitive stuff
> of the world* (tohu va-vohu) *was called
> into being from nothingness, all divine
> activity was restricted to the production of
> new forms and structures and combina-
> tions from pre-existent material; in the be-
> ginning there was "creation," beriah (i.e.,
> out of nothing), but thereafter came only
> "formation," yetzirah (i.e., out of previ-
> ous stuff). Life is no exception to this rule:
> it, too, was formed from material that ex-
> isted before it, since the moment of cre-
> ation. Thus, vegetation was brought out
> from the earth (Gen. 1:11), fish from the
> water (Gen. 1:20), animals from the earth
> (Gen. 1:24), etc. Even man was created
> out of dust from the ground (Gen. 2:7). In
> each of these cases, the Torah implicitly
> grants that natural chemical and biologi-
> cal processes were utilized by the Creator*

to produce His creations. Man, too, inso-
far as he is a natural being, was the result
of a natural developmental process. (The
only difference is in a realm other than
the natural: man is also a metaphysical
being, he represents an interpenetration
of the material and the divine.) The cre-
ation of life is, therefore, according to the
Bible, no more and no less "miraculous"
than the creation of any of the complex in-
organic substances that were formed out of
the primordial chaos after the first instant
of creatio ex nihilo.[10]

Theological concern with evolution tends to be derived from a literalist reading of Genesis. As I have tried to show, in Jewish religious thought Genesis is not regarded as meant for a literal reading, and Jewish tradition has not usually read it so. The theological issue in Judaism turns on the question of what one thinks God intended to reveal in Genesis, i.e., whether it is meant as a blueprint of creation, or as a more abstract truth, namely, that the world is not a random surd but the result of Divine Concern and Purpose. This problem of purpose is the central theological issue created by the Darwinian theory. The basis for disagreement is not the conflict of evolution with a literal reading of Genesis, but rather the evolutionist's denial of teleology, i.e., the denial of purpose in and through nature, and purposeful movement in and through history, toward some end or goal.

While evolution argues for the random, purposeless nature of natural selection, this argument only describes specific events, whether mutations or reproductions, *within* history and nature. It does not offer evidence for or against the purposeful ordering of nature and history as wholes. As the medievals, for example, Thomas Aquinas in his Five-Fold Way, were wrong when they argued that the existence of a "First Cause" could be proven inductively as a consequence of observing chain of causation within nature and history, because the observance of a cause *in* nature or history does not prove there is a cause *of* nature or history; so, too, modern men, who deny a cause of nature because of the randomness of natural selection, make the same error of logic, but in the reverse direction. For modern evolutionists take the apparent absence of a cause *in* any given event or chain of events as grounds for eliminating the possibility of a cause *of* the entire process.

The true radicalism of modern science resides in its denial of teleological causation. We must, however, recognize that teleology is a metaphysical concept whose ultimate reality cannot be affirmed or denied on the basis of empirical or scientific evidence. Despite scientific claims to the contrary, the destiny or meaning of the human race, and of the cosmic order, cannot be ascertained by a study of discrete biological or historical events. It is no more logical to argue that the world has no ultimate cause or purpose than to argue that it does—in both cases the empirical or scientific evidence for deciding the matter is inadequate.

NOTES

1. Judaism refers here to rabbinic or orthodox Judaism, representing the mainstream of historic Jewish religious belief and practice.

2. Torah, best translated as "teaching," i.e., Divine teaching or instruction, rather than "Law", is used in several senses. The most precise refers to the Five Books of Moses. The more extended includes the whole Hebrew Bible.

3. The Mishna is the earliest stratum of rabbinic legal discussion which was edited and "closed" about 200 C.E. in Palestine. The Gemara is the second, more developed stratum of Jewish law which was edited twice, once in Palestine in the 5th century C.E., known as the Palestinian Talmud or Jerusalem Talmud, and once in the 6th century C.E. in Babylonia, known as the Babylonian Talmud. See the *Encyclopedia Judaica* (Jerusalem, 1972) for more extended discussion.

4. S. Katz, *Jewish Ideas and Concepts* (New York: Schocken, 1978).

5. Maimonides, *Guide for the Perplexed*, 1:1.

6. *Bereshit Rabba (Midrash on Genesis)*, 1:9.

7. Maimonides, *Guide for the Perplexed*, 2:16.

8. For a fuller explanation and discussion of the concepts *aggadah* (*aggadic*) and *halachah* (*halachic*), see the entries in the *Encyclopedia Judaica*.

9. *Fathers According to Rabbi Nathan*, translated by J. Goldin (New Haven: Yale University Press, 1955), ch. 1, p. 11.

10. Norman Lamm, *Extraterrestrial Life* in *Challenge*, ed. A. Carmell and C. Domb (New York: Feldheim Publishing Company, 1976), pp. 383–384.

Sources

The material for the biography of Einstein was drawn primarily from the following books: *Albert Einstein, Creator and Rebel*, by Banesh Hoffmann and Helen Dukas, (New York: Viking Press, 1972); *Einstein*, by Jeremy Bernstein, (New York: Viking Press, 1973); *Einstein, His Life and Times*, by Philipp Frank, (New York: Alfred A. Knopf, 1967). The books by Bernstein and by Hoffmann and Dukas are particularly clear and readable accounts of Einstein's scientific work.

SOURCES

The Einstein letters to de Sitter were obtained from Leiden Observatory through the courtesy of Professor H. C. Van der Hulst. The statements of Einstein's religious views were obtained from *Ideas and Opinions*, by Albert Einstein, (New York: Crown Publishers, 1973). The quotation on page 38 came from *Forty Minutes with Einstein* by A. V. Douglas (Journal of the Royal Astronomical Society of Canada, volume 50, page 100, 1956). The description of Einstein's relationship to Friedmann in Chapter 2 was based on George Gamow's account in his autobiography, *My World Line*, (New York: Viking Press, 1970), on a letter from Friedmann to Einstein provided through the courtesy of Helen Dukas of the Institute for Advanced Study in Princeton, and on Einstein's publications in the *Zeitschrift fur Physik*, (volume 11, p. 326, 1922 and volume 16, p. 228, 1923).

The biographical information on Hubble was taken from an article by Nicholas Mayall in *Biographical Memoirs, Volume 51, National Academy of Sciences*, (New York: Columbia University Press, 1970). The remarks by Hubble on the work of Slipher and Humason are taken from the published version of the George Darwin Lecture, *The Law of Red-Shifts*, delivered on May 8, 1953. A complete account of the developments leading to the formulation of the Hubble Law can be found in the very interesting Ph.D. thesis, *"The Velocity-*

Distance Relation" by Norriss Swigart Hethering-
ton, (Department of History and Philosophy of
Science, Indiana University, 1970).

The material on the life of Humason was tak-
en partly from an article in the *Quarterly Journal*
of the Royal Astronomical Society, volume 14, p.
235 (1973). Dr. John Hall provided additional information regarding Humason's earliest years at
Mount Wilson, and also regarding the circumstances involved in Slipher's research on the spiral
nebulas.

I am indebted to Mrs. Theodora Smit, Willem de Sitter's daughter, for the recollections of
her family life and her father's early years in Cape
Town. The remainder of the biographical information on de Sitter was drawn from *Man Discovers
the Galaxies*, by Richard Berendzen, Richard Hart
and Daniel Seeley, (New York: Science History
Publications, 1976). I am also indebted to Drs.
Ralph Alpher and Robert Herman for information
on the circumstances of their early work.

The following sources provided materials for
the discussion in Chapter 6 and the Epilogue on
the reactions of scientists to new theories of the
Universe: *The Expanding Universe*, by Arthur Eddington, (Cambridge: The University Press,
1952); *The Nature of the Physical World*, by Arthur Eddington, (Cambridge: The University
Press, 1953); *Modern Cosmology and the Christian
Idea of God*, by E. A. Milne, (Oxford: The Claren-

SOURCES

don Press, 1952); *The Beginning and End of the World*, by Edmund Taylor Whittaker, (London: Humphrey Milford, 1952); Phillip Morrison, transcript of the BBC film, *The Violent Universe*; Allan Sandage, *Time*, December 30, 1974; P. J. E. Peebles, Transcript of Nova program, *A Whisper from Space*.

Picture Credits

Page	
6	Courtesy Ralph Alpher
7	AIP, Niels Bohr Library
8–9 *and* 10–11	Courtesy Bell Laboratories
19	Courtesy Frank Edmondson
21	AIP, Niels Bohr Library
22–23	American Astronomical Society
24–25	Yerkes Observatory
26–27 *and* 28	Courtesy Theodora Smit
31	AIP, Niels Bohr Library
32–33	Courtesy Theodora Smit
34–35	Wide World Photos
44–45	AIP, Niels Bohr Library
47 *and* 48–49	Lick Observatory
50–51	Courtesy Margaret Harwood
52–53	Scientific American
54–55	Hale Observatories

PICTURE CREDITS

Index

Alpher, Ralph, 5, *6*, 169
Adams, Walter, *65*
American Astronomical Society,
    13, 22
Andromeda galaxy (nebula), 20,
    46–49, 87, 92–93
Antimatter, 116
Atoms,
    in stars, 96
    in the Universe, 118
Augustine, St., 102, 105

Bell Laboratories, 4

Big Bang, 2, 110
    evidence for, 4–5
    objections to, 103–104
Cepheid star, 41, 45
Collapsing universe, 108–109
Coma cluster, 112
Cosmic explosion, 3
Cosmic fireball radiation
    detection of, 4–5, 8, 10
    evidence against Steady State
        Theory, 99
    prediction of, 6

INDEX

Cosmos: See Universe

Density of matter in the Universe, 110–113

de Sitter, Eleonora Suermondt, *28*

de Sitter, Willem,
  correspondence with Einstein, 17, 32
  life of, 27–28
  prediction of expanding Universe, 14, 16–17, 29

de Sitter Universe, 29

Dukas, Helen, *80*, 165

Earth, movement through the galaxy, 37

Eddington, Arthur, 17, 29, *31*, 43, 102

Ehrenfest, Paul, 30, *31*

Einstein, Albert,
  and de Sitter, 14, 17, 29, *31*, *32–33*
  life of, 72–83
  objections to the theory of the expanding Universe, 17, 29, 32–33, 43, 103–104

Expanding Universe, 3
  de Sitter's prediction of, 14, 29
  Hubble's Law and, 42, 50, 86
  objections to, 102–103
  Slipher's observation of, 13–15, 29

Frank, Philipp, 73, 74

Friedmann, Alexander,
  and Einstein, 15–16
  model of expanding Universe, 43

Galaxies
  clusters of, 112
  expanding motion of, 3, 14
  nature of, 40
  speeds of and distances to, 32–42

Gamow, George, 6, *7*

Great Plan, 18

Hall, John, 14, 167

Herman, Robert, 5, *6–7*

Horn antenna, *8–9*, *10*–11

Hubble, Edwin P.,
  and Slipher, 14, 17, 37–38
  life of, *60–69*
  studies of galaxies, 37–42

Hubble's Law, 42, 45, 85–86, 91

Hubble, Lucy, *63*

Humason, Milton,
  measurement of speeds of galaxies, 38–39, 50–*53*

Hydrogen,
  abundance in Cosmos, 95–96, 100
  depletion in Universe, 99

Island Universe: See galaxy

Jeans, James, 40, *65*, *66–67*

Kapteyn, Jacobus, *24*

Kapteyn Universe, 24–25

Lemaître, Georges, 34–35, 43

Lorentz, Hendrick Antoon, 29–30, 31

Lowell Observatory, 14, 20

Miller, John, 13–14

Milne, Edward A., 102, 115, 167–168

Morrison, Phillip, 103, 168

Mount Wilson Observatory, 37–38, *54–55*, 61, *65*

Nernst, Walter, 103

Newton, Isaac, 73

Nuclear reactions, 117

Oscillating universe, 109–110

Palomar Mountain, *56*, 61, 103

Peebles, James, 115–116, 167

Penzias, Arno, 4–5, *8–9*, *10–11*

Quasars, 119

Red shift, 14, 43
  measurements of, 50–51, *52–53* 87–90

Relativity,
  general, 14–15, 29, 42, 73
  special, 29

Sandage, Allan, 94fn., 103, 168

Slipher, Vesto Melvin,
  discovery of outward-moving galaxies, 13–14, *19*, 20, 22, 37
  and Kapteyn, 24
  and Hubble, 14, 17, 37–38

Smit, Theodora, 28, 167

Spectrum, 88, 89

Spinoza, 17

Spiral galaxies (nebulas), 20, 39–41, 45, 54

Stars,
  birth of, 95–97
  death of, 97–99
  production of elements in, 97

Steady State cosmology, 4, 94fn., 99fn.
  evidence against, 5

Sun,
  location in galaxy, 24–25, 37

Supernova, 98, *127*

Thermodynamics, Laws of, 101

Universe,
  age of, 3
  beginning of, 1–5
  collapsing, 108–109
  de Sitter's, 29
  first billion years of, 115–120
  island universes, 39, 40, 42, 45, 46, 54
  oscillating, 109–110

Whitehead, Alfred, 73–74

Whittaker, Edmund, 101–102

Wilson, Mount: See Mount Wilson Observatory

Wilson, Robert, 4, 5, *8–9*, *10–11*

Yerkes Observatory, 24